P9-AFS-938

THE PIANO BOOK

THE PIANO BOOK

Buying & Owning a New or Used Piano

Second Edition

Larry Fine

Illustrated by Douglas R. Gilbert

Foreword by Keith Jarrett

BROOKSIDE PRESS • BOSTON, MASSACHUSETTS

Brookside Press
P.O. Box 178, Jamaica Plain, Massachusetts 02130
(617) 522-7182
(800) 545-2022 (orders)

The Piano Book copyright © 1987, 1990 by Lawrence Fine
All rights reserved

"Foreword" copyright © 1987 by Keith Jarrett
All rights reserved

Printed in the United States of America

No part of this book may be reproduced in any form whatsoever
without prior written permission of the publisher, except for brief
quotations embodied in critical articles and reviews.

Cover design by Laurie Dolphin

Library of Congress Catalog Card Number 89-81394

ISBN 0-9617512-3-1
ISBN 0-9617512-2-3 (pbk.)

Dimensions Math
Workbook 2B

W9-CUB-699

Authors and Reviewers

Jenny Kempe

Bill Jackson

Tricia Salerno

Allison Coates

Cassandra Turner

Singapore Math Inc.

Published by Singapore Math Inc.

19535 SW 129th Avenue

Tualatin, OR 97062

www.singaporemath.com

Dimensions Math® Workbook 2B

ISBN 978-1-947226-21-0

First published 2018

Reprinted 2019, 2020 (three times), 2021 (three times)

Copyright © 2017 by Singapore Math Inc.

All rights reserved. This book or any portion thereof may not be
reproduced or used in any manner whatsoever without the express
written permission of the publisher.

Printed in China

Acknowledgments

Editing by the Singapore Math Inc. team.

Design and illustration by Cameron Wray with Carli Bartlett.

Contents

Chapter	Exercise	Page

Chapter	Exercise	Page

Chapter	Exercise	Page

This workbook includes **Basics** (a review of basic concepts) and problems for **Practice**, for **Challenge**, and to **Check** understanding.

Chapter 8 Mental Calculation

Basics

1 (a) $48 + 6 = 50 + \boxed{}$

48 ──○── 4

$48 + 6 = \boxed{}$

(b) $348 + 6 = \boxed{} + 4$

348 ──○── 4

$348 + 6 = \boxed{}$

---

2 (a) $86 + 9 = \boxed{} + 10$

86 ──○── 1

$86 + 9 = \boxed{}$

(b) $686 + 9 = \boxed{} + 10$

686 ──○── 1

$686 + 9 = \boxed{}$

---

3 (a) $67 + 7 = 60 + \boxed{}$

67 ──60──○

$67 + 7 = \boxed{}$

(b) $467 + 7 = 460 + \boxed{}$

467 ──460──○

$467 + 7 = \boxed{}$

---

4 $539 + 7 = 540 + \boxed{} = \boxed{}$

Practice

5 Add.

67 + 8 = [ ] **S**	49 + 6 = [ ] **I**	9 + 37 = [ ] **A**
632 + 5 = [ ] **T**	8 + 483 = [ ] **M**	324 + 7 = [ ] **L**
8 + 208 = [ ] **P**	444 + 7 = [ ] **N**	784 + 9 = [ ] **H**
617 + 5 = [ ] **E**	7 + 533 = [ ] **D**	425 + 6 = [ ] **B**
142 + 8 = [ ] **R**	193 + 7 = [ ] **Y**	5 + 898 = [ ] **O**

Joke: Why did the math book look so sad?

Write the letters that match the answers above to find out.

492	55	637	341	793	46	540	504	637	903	903	638	218

491	46	451	200	56	216	150	903	431	331	622	491	75

Basics

1 (a) $60 + 70 = 100 +$ ☐

40   ○

$60 + 70 =$ ☐

(b) $360 + 70 =$ ☐ $+ 30$

○   30

$360 + 70 =$ ☐

2 (a) $240 + 80 =$ ☐ $+ 100$

○   20

$240 + 80 =$ ☐

(b) $247 + 80 =$ ☐ $+ 100$

○   20

$247 + 80 =$ ☐

3 (a) $650 + 60 =$ ☐ $+ 100$

○   40

$650 + 60 =$ ☐

(b) $652 + 60 =$ ☐ $+ 100$

○   40

$652 + 60 =$ ☐

4 $539 + 90 = 529 +$ ☐ $=$ ☐

Practice

5 Add.

90 + 50 = ☐ **S**	72 + 80 = ☐ **A**	570 + 60 = ☐ **D**
70 + 150 = ☐ **T**	850 + 50 = ☐ **R**	620 + 90 = ☐ **B**
483 + 90 = ☐ **R**	387 + 70 = ☐ **I**	30 + 291 = ☐ **A**
796 + 80 = ☐ **N**	60 + 552 = ☐ **B**	334 + 40 = ☐ **O**
80 + 155 = ☐ **R**	248 + 60 = ☐ **P**	399 + 90 = ☐ **S**

What two animals can see behind themselves without turning their heads?
Write the letters that match the answers above to find out.

308	321	573	900	374	220	140	603	152	876	630

866	720	235	321	710	612	457	220	489	335	800

Basics

1 (a) 9 tens ☐ ones = 100

(b) ☐ tens 5 ones + 5 tens ☐ ones = 9 tens 10 ones

(c) 45 + ☐ = 100

2 (a) 60 + ☐ = 100   100 − 60 = ☐

(b) 64 + ☐ = 100   100 − 64 = ☐

(c) 64 + ☐ = 300   300 − 64 = ☐

3 Match pairs of numbers that make 100.

84   96

49   78

73   85

4   16

22   51

15   27

Practice

1 Write the missing numbers.

(a) $74 + \boxed{} = 100$ (b) $15 + \boxed{} = 100$

(c) $55 + \boxed{} = 100$ (d) $81 + \boxed{} = 100$

(e) $92 + \boxed{} = 100$ (f) $\boxed{} + 22 = 100$

(g) $100 = \boxed{} + 68$ (h) $100 = 3 + \boxed{}$

2 Subtract.

(a) $100 - 46 = \boxed{}$ (b) $100 - 62 = \boxed{}$

(c) $100 - 33 = \boxed{}$ (d) $100 - 57 = \boxed{}$

(e) $100 - 75 = \boxed{}$ (f) $\boxed{} = 100 - 6$

(g) $100 - \boxed{} = 98$ (h) $64 = 100 - \boxed{}$

Challenge

6 The numbers on each side of each triangle should add up to 100. Write the missing numbers.

(a)

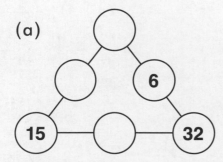

(b)

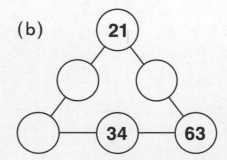

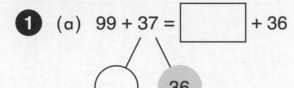

Exercise 4

Basics

1 (a) $99 + 37 = \boxed{} + 36$

           ○   **36**

    $99 + 37 = \boxed{}$

(b) $99 + 537 = 100 + \boxed{}$

      1   ○

    $99 + 537 = \boxed{}$

---

2 (a) $85 + 98 = \boxed{} + 100$

      ○   **2**

    $85 + 98 = \boxed{}$

(b) $485 + 98 = \boxed{} + 100$

      ○   **2**

    $485 + 98 = \boxed{}$

---

3 $73 \xrightarrow{\;+\,100\;} \boxed{} \xrightarrow{\;-\,1\;} \boxed{}$

   $73 \xrightarrow{\;+\,99\;} \boxed{}$

   $73 + 99 = \boxed{}$

4 (a) $97 = 100 - \boxed{}$

(b) $97 + 400 = 500 - \boxed{} = \boxed{}$

(c) $97 + 486 = 586 - \boxed{} = \boxed{}$

5 Add.

54 + 99 = ☐	38 + 97 = ☐	73 + 98 = ☐
98 + 97 = ☐	99 + 88 = ☐	142 + 98 = ☐
652 + 99 = ☐	325 + 98 = ☐	97 + 606 = ☐
777 + 98 = ☐	98 + 709 = ☐	447 + 97 = ☐
97 + 119 = ☐	324 + 98 = ☐	99 + 599 = ☐

Color the spaces that contain the answers to help the spider find its home.

204	671	195	153	807	124	806
117	175	135	421	751	432	423
874	875	187	545	544	689	171
216	422	702	973	240	703	698
578	571	214	230	402	161	398

Check

1 Complete the number patterns.

(a)

138		152	159			180

(b)

230	310	390				

(c)

333	432					927

(d)

147	244					729

2 Find the pattern.
Write the missing numbers.

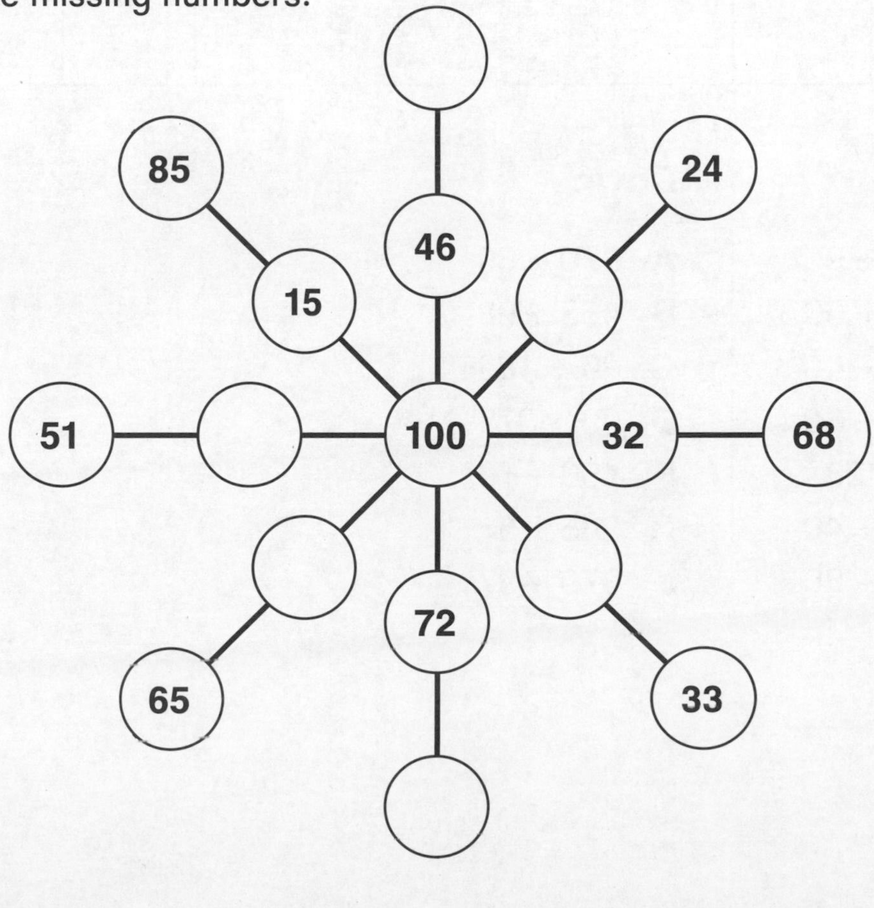

(3) Complete the cross-number puzzle using the clues.

Across

A 936 + 9
D 360 + 60
E 619 + 40
G 373 + 50
I 6 + 728
K 660 + 40
L 580 + 80

Down

A 870 + 80
B 468 + 98
C 90 + 189
D 99 + 375
F 453 + 70
H 245 + 5
J 379 + 97

4 June is collecting minerals to polish in a rock tumbler.
She has collected 90 rose quartz.
She has 40 more smoky quartz than rose quartz.

(a) How many smoky quartz does she have?

She has _____ smoky quartz.

(b) How many of both kinds of quartz does she have in all?

She has _____ quartz in all.

5 June had 98 moonstones, and then found another
38 moonstones and 80 sunstones.

(a) How many moonstones does she have in all?

She has _____ moonstones in all.

(b) How many moonstones and sunstones does she have in all?

She has _____ moonstones and sunstones in all.

Challenge

6 Complete the number pattern.

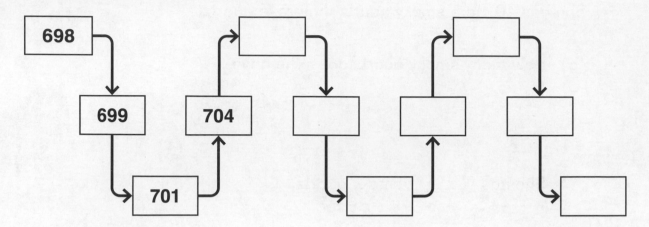

7 Add using mental math.

(a) 63 + 499 = ☐

(b) 24 + 598 = ☐

(c) 99 + 799 = ☐

(d) 398 + 265 = ☐

8 Mei added 85 and 67 like this:

85 + 67 = 85 + 70 − 3 = 152

Use this strategy to mentally add the following.

(a) 437 + 8 = ☐

(b) 95 + 78 = ☐

(c) 59 + 35 = ☐

(d) 388 + 68 = ☐

Basics

1 (a) 82 − 6 = [ ] + 74

(circle) — 80

82 − 6 = [ ]

(b) 382 − 6 = [ ] + 74

(circle) — 80

382 − 6 = [ ]

2 (a) 86 − 7 = [ ] + 3

(circle) — 10

86 − 7 = [ ]

(b) 286 − 7 = [ ] + 3

(circle) — 10

286 − 7 = [ ]

3 (a) 51 − 5 = [ ] − 4

(circle) — 4

51 − 5 = [ ]

(b) 351 − 5 = [ ] − 4

(circle) — 4

351 − 5 = [ ]

4 (a) 62 − 8 = [ ] + 4

(circle) — 12

62 − 8 = [ ]

(b) 762 − 8 = [ ] + 4

(circle) — 12

762 − 8 = [ ]

Practice

 5 Subtract.

140 − 90 = ☐ **Y**	330 − 80 = ☐ **A**	501 − 20 = ☐ **D**
470 − 90 = ☐ **S**	290 − 30 = ☐ **O**	630 − 60 = ☐ **R**
330 − 50 = ☐ **U**	740 − 70 = ☐ **E**	210 − 90 = ☐ **T**
881 − 50 = ☐ **M**	965 − 90 = ☐ **L**	108 − 10 = ☐ **N**
487 − 90 = ☐ **K**	512 − 50 = ☐ **C**	209 − 60 = ☐ **E**

Riddle: What always holds you up when you are late?

Write the letters that match the answers above to find out.

50	260	280	570	240	831	280	380	462	875	149	380

250	98	481	931	380	397	670	875	149	120	260	98

Basics

1 (a) $300 - 99 = 200 +$ ▢

200 ○

$300 - 99 =$ ▢

(b) $305 - 99 =$ ▢ $+ 1$

○ 100

$305 - 99 =$ ▢

2 (a) $500 - 98 = 400 +$ ▢

400 ○

$500 - 98 =$ ▢

(b) $585 - 98 =$ ▢ $+ 2$

○ 100

$585 - 98 =$ ▢

3 $173 \xrightarrow{-100}$ ▢ $\xrightarrow{+1}$ ▢

$173 \xrightarrow{-99}$ ▢

$173 - 99 =$ ▢

4 (a) $100 = 97 +$ ▢

(b) $400 - 97 = 300 +$ ▢ $=$ ▢

(c) $486 - 97 = 386 +$ ▢ $=$ ▢

Practice

5 Subtract.

154 − 99 = ☐	600 − 97 = ☐	173 − 98 = ☐
796 − 97 = ☐	301 − 99 = ☐	142 − 98 = ☐
652 − 99 = ☐	325 − 98 = ☐	891 − 97 = ☐
777 − 98 = ☐	313 − 97 = ☐	443 − 98 = ☐
734 − 97 = ☐	324 − 98 = ☐	998 − 99 = ☐

Riddle: What is at the end of a rainbow?

Color the spaces that contain the answers to find out.

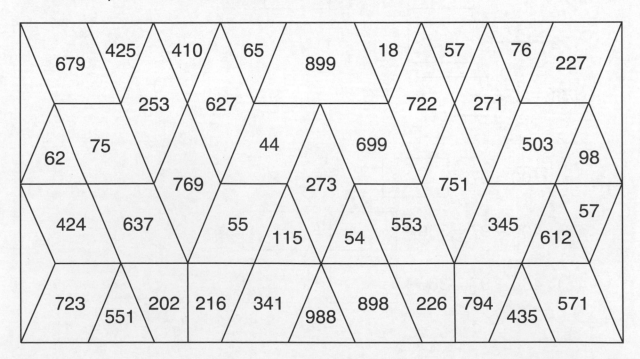

Check

 Subtract.

410 − 70 = ☐ **U**	500 − 98 = ☐ **A**	100 − 62 = ☐ **Y**
800 − 30 = ☐ **T**	120 − 50 = ☐ **M**	632 − 7 = ☐ **E**
546 − 50 = ☐ **H**	740 − 97 = ☐ **C**	210 − 90 = ☐ **O**
333 − 40 = ☐ **E**	525 − 9 = ☐ **N**	632 − 98 = ☐ **R**

Joke: What did one wall say to the other?

Write the letters that match the answers above to find out.

70	625	293	770	596	38	120	340	210	402	770	562

48	770	496	625	637	643	120	534	516	293	534	435

2 Complete the cross-number puzzle using the clues.

Across

A 741 – 60
D 733 – 7
G 137 – 98
I 620 – 40
J 91 – 9
M 495 – 60
O 712 – 98
Q 830 – 70
R 268 – 20

Down

B 100 – 17
C 116 – 97
D 760 – 8
E 100 – 72
F 614 – 5
K 151 – 98
L 840 – 80
N 582 – 4
P 506 – 30

3 June made 132 earrings and 80 necklaces last month using polished rocks.

(a) How many more earrings than necklaces did she make?

She made _____ more earrings than necklaces.

(b) June sold 98 of the earrings.
How many earrings did she have left?

She has _____ earrings left.

4 June made $350 from selling jewelry.
She wants to make $410.

(a) How much more money does she need to make?

She needs to make $_____ more.

(b) She spent $90 of the $350 she made on more craft items to make more jewelry.
How much money does she have left of that $350?

She has $_____ left.

Challenge

5 Complete the number pattern.

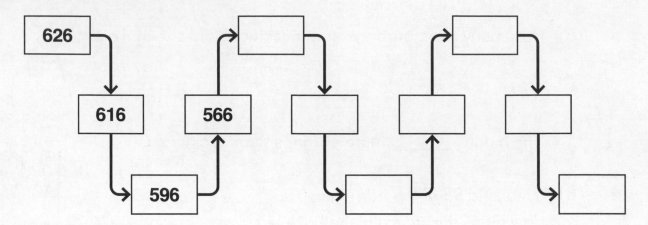

6 Subtract using mental math.

(a) $245 - 199 =$ ☐

(b) $963 - 499 =$ ☐

(c) $391 - 297 =$ ☐

(d) $924 - 598 =$ ☐

7 Dion subtracted 68 from 453 like this:

$453 - 68 = 453 - 70 + 2 = 385$

Use the strategy to mentally subtract the following.

(a) $437 - 8 =$ ☐

(b) $46 - 29 =$ ☐

(c) $59 - 35 =$ ☐

(d) $324 - 48 =$ ☐

Check

1 Find the values.

(a) 237 + 98 = ☐   (b) 237 − 98 = ☐

(c) 64 + 99 = ☐   (d) 132 − 98 = ☐

(e) 373 − 97 = ☐   (f) 425 + 98 = ☐

(g) 278 − 99 = ☐   (h) 631 + 97 = ☐

(i) 843 + 99 = ☐   (j) 555 − 98 = ☐

2 Write >, <, or = in each ◯.

(a) 100 − 65 ◯ 100 − 56

(b) 42 + 612 ◯ 642 + 12

(c) 82 + 60 ◯ 142 − 50

(d) 147 − 99 ◯ 47 + 99

(e) 375 + 8 ◯ 317 + 80

(f) 630 − 80 ◯ 480 + 70

(g) 132 + 7 + 40 ◯ 124 + 7 + 30

(h) 800 + 40 + 50 ◯ 900 − 90

3 Follow the arrows and fill in the missing numbers.

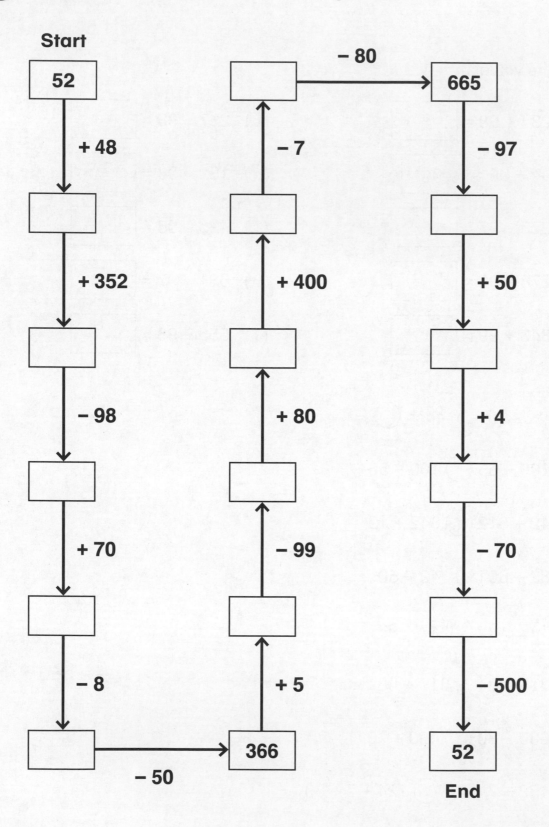

Start

52 → +48 → 100 → +352 → 452 → −98 → 354 → +70 → 424 → −8 → 416 → −50 → 366

366 → +5 → 371 → −99 → 272 → +80 → 352 → +400 → 752 → −7 → 745 → −80 → 665

665 → −97 → 568 → +50 → 618 → +4 → 622 → −70 → 552 → −500 → 52

End

4 There are 206 bones in an adult human skeleton.
30 of them are in the spine.
How many bones are not in the spine?

5 Dexter found a snake skeleton and counted 310 vertebrae.
Each vertebra had 2 ribs attached.

(a) How many ribs did the snake skeleton have?

(b) How many bones did it have for both vertebrae and ribs?

(c) The head of the snake had 10 bones.
How many bones did the snake have in all?

(d) How many more bones did this
snake have than an adult human skeleton?

Challenge

6 Find the number that each shape stands for.

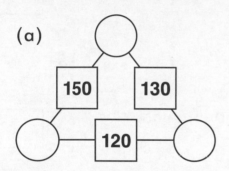

◆ + ⬡ = 100

◆ − ⬡ = 26

26 + ◆ = 89

◆ = [ ]

⬡ = [ ]

7 The numbers on each circle should add up to the number in the square between them.
Write the missing numbers.

(a)

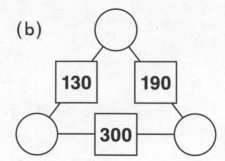

150   130

120

(b)

130   190

300

⋮ There are some bees and flowers.
If each bee lands on a different flower, one bee does not get a flower.
If two bees share each flower, there is one flower left out.
How many flowers and bees are there?

Chapter 9 Multiplication and Division of 3 and 4

Basics

1 Count by threes and complete the multiplication equations.

$1 \times 3 =$ ☐

$2 \times 3 =$ ☐

$3 \times 3 =$ ☐

$4 \times 3 =$ ☐

$5 \times 3 =$ ☐

$6 \times 3 =$ ☐

$7 \times 3 =$ ☐

$8 \times 3 =$ ☐

$9 \times 3 =$ ☐

$10 \times 3 =$ ☐

2 The sum of the digits in the products is _____, _____, or _____.

Practice

3 (a) 3 × 3 is _____ more than 2 × 3.

3 × 3 = [ ]

(b) 4 × 3 is 3 more than _____ × 3.

4 × 3 = [ ]

(c) 5 × 3 = [ ]

6 × 3 = 15 + [ ] = [ ]

7 × 3 = 15 + [ ] = [ ]

(d) 9 × 3 = 30 − [ ] = [ ]

8 × 3 = [ ] − 6 = [ ]

4 Each cake has 3 candles.
How many candles are on 7 cakes?

[ ] × [ ] = [ ]

_____ candles are on 7 cakes.

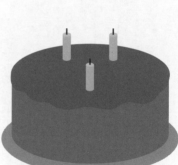

5 Circle products of 3.

17   16   12   9   25   21   18

Basics

1

$$3 + 3 + 3 + 3 + 3 + 3 = \boxed{}$$ $\qquad$ $6 + 6 + 6 = \boxed{}$

$$6 \times 3 = \boxed{}$$ $\qquad\qquad\qquad$ $3 \times 6 = \boxed{}$

2 $8 \times 3 = \boxed{}$

$3 \times 8 = \boxed{}$

3 (a) $\quad 5 \times 3 = \boxed{}$ $\qquad$ $3 \times 5 = \boxed{}$

(b) $\quad 2 \times 3 = \boxed{}$ $\qquad$ $3 \times 2 = \boxed{}$

(c) $\quad 7 \times 3 = \boxed{}$ $\qquad$ $3 \times 7 = \boxed{}$

(d) $10 \times 3 = \boxed{}$ $\qquad$ $3 \times 10 = \boxed{}$

(e) $\quad 1 \times 3 = \boxed{}$ $\qquad$ $3 \times 1 = \boxed{}$

(f) $\quad 9 \times 3 = \boxed{}$ $\qquad$ $3 \times 9 = \boxed{}$

(g) $\quad 4 \times 3 = \boxed{}$ $\qquad$ $3 \times 4 = \boxed{}$

(h) $\quad 3 \times 3 = \boxed{}$

Practice

4 Match.

3 × 3	9	3 × 2
3 × 6	18	3 × 5
10 × 3	30	4 × 3
3 × 7	3	3 × 10
2 × 3	12	8 × 3
3 × 8	6	6 × 3
3 × 4	24	3 × 3
9 × 3	15	7 × 3
5 × 3	21	3 × 9
1 × 3	27	3 × 1

Basics

1 Dion is planting 3 tomato seeds in each jiffy pot.
How many jiffy pots does he need for 24 seeds?

☐ × 3 = 24 | _____ groups of 3 is 24.

24 ÷ 3 = ☐

He needs _____ jiffy pots.

2 Sofia divided 18 jiffy pots equally onto 3 trays.
How many pots did she put on each tray?

3 × ☐ = 18 | 3 groups of _____ is 18.

18 ÷ 3 = ☐

She put _____ jiffy pots on each tray.

Practice

 3

$\boxed{} \times 3 = 12$

$12 \div 3 = \boxed{}$

$\boxed{} \times 3 = 9$

$9 \div 3 = \boxed{}$

$\boxed{} \times 3 = 30$

$30 \div 3 = \boxed{}$

$\boxed{} \times 3 = 15$

$15 \div 3 = \boxed{}$

$\boxed{} \times 3 = 24$

$24 \div 3 = \boxed{}$

$\boxed{} \times 3 = 3$

$3 \div 3 = \boxed{}$

$\boxed{} \times 3 = 18$

$18 \div 3 = \boxed{}$

$\boxed{} \times 3 = 6$

$6 \div 3 = \boxed{}$

$\boxed{} \times 3 = 21$

$21 \div 3 = \boxed{}$

$\boxed{} \times 3 = 27$

$27 \div 3 = \boxed{}$

4 (a) $\boxed{} \div 3 = 4$

(b) $\boxed{} \div 3 = 9$

(c) $\boxed{} \div 3 = 7$

(d) $\boxed{} \div 3 = 5$

(e) $\boxed{} \div 3 = 6$

(f) $\boxed{} \div 3 = 8$

5 Avery, Dana, and Grace share a box of 12 colored pencils equally.
How many pencils does each girl get?

$12 \div 3 =$ ⬜

Each girl gets _____ pencils.

6 There are 21 tennis balls.
Sharif puts 3 tennis balls in each can.
How many cans does he need?

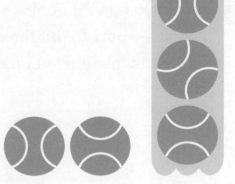

⬜ ◯ ⬜ = ⬜

He needs _____ cans.

7 Laila has a ribbon that is 9 feet long.
She cuts it into 3 equal pieces.
How long is each piece?

⬜ ◯ ⬜ ◯ ⬜

Each piece is _____ feet long.

Challenge

8 Caleb has 2 boxes with 6 markers in each box.
He and two friends share them equally.
How many markers does each boy get?

Each boy gets _____ markers.

9 Mariya has 20 photos and 3 pages in her album.
If she wants to put the same number of photos on each page,
what is the least number of photos she will have left over?

She will have _____ photos left over.

10 Ryan is placing stakes 3 inches apart from each other.
The distance from the first to the last stake is 30 inches.
How many stakes has he placed?

.3 in.

He has placed _____ stakes.

Check

1 Add or subtract.

$896 + 48 = \boxed{}$ | $279 + 107 = \boxed{}$ | $456 + 365 = \boxed{}$

$148 - 82 = \boxed{}$ | $432 - 63 = \boxed{}$ | $803 - 29 = \boxed{}$

2 Write $+$, $-$, $\times$, $\div$, or $=$ in each $\bigcirc$.

(a) $15 \bigcirc 3 \bigcirc 5$ (b) $15 \bigcirc 3 \bigcirc 12$

(c) $15 \bigcirc 3 \bigcirc 18$ (d) $12 \bigcirc 3 \bigcirc 9$

(e) $24 \bigcirc 3 \bigcirc 21$ (f) $21 \bigcirc 3 \bigcirc 7$

(g) $7 \bigcirc 3 \bigcirc 10$ (h) $10 \bigcirc 3 \bigcirc 30$

(i) $3 \bigcirc 3 \bigcirc 1$ (j) $3 \bigcirc 3 \bigcirc 6$

3 Multiply or divide.

$2 \times 5 = \boxed{}$ **S**	$9 \times 5 = \boxed{}$ **N**	$6 \times 3 = \boxed{}$ **H**
$10 \times 3 = \boxed{}$ **R**	$15 \div 3 = \boxed{}$ **T**	$27 \div 3 = \boxed{}$ **U**
$7 \times 2 = \boxed{}$ **R**	$10 \div 5 = \boxed{}$ **E**	$3 \times 2 = \boxed{}$ **S**
$4 \times 3 = \boxed{}$ **N**	$3 \times 7 = \boxed{}$ **E**	$18 \div 1 = \boxed{}$ **H**
$24 \div 3 = \boxed{}$ **R**	$9 \div 3 = \boxed{}$ **I**	$8 \times 2 = \boxed{}$ **E**
$35 \div 5 = \boxed{}$ **O**	$5 \times 10 = \boxed{}$ **T**	$20 \div 5 = \boxed{}$ **O**

Write the letters that match the answers above to learn a fun fact.

18	4	8	10	21	6	15	14	9	12	24	7	45

11	50	18	16	3	30	27	5	4	2	6	19	36

4 3 bags of flour weigh 9 kilograms.
How much does one bag of flour weigh?

One bag of flour weighs _____ kg.

5 A bag of potatoes weighs 3 kilograms.
How much do 9 bags of potatoes weigh?

9 bags of potatoes weigh _____ kg.

6 Arman can fit 15 pots equally onto 3 trays.

(a) How many pots go on each tray?

_____ pots go on each tray.

(b) How many trays are needed for 25 pots?

_____ trays are needed for 25 pots.

Challenge

7 ◆ + ◆ + ◆ = 12

◆ + ◆ + ● + ● = 14

● + ● + ● = [　　　]

8 There are 10 tricycles and bicycles in all.
It there are 23 wheels, how many are bicycles and how many are tricycles?

There are _____ bicycles and _____ tricycles.

9 A piece of string is 20 ft long.
It is cut into as many pieces as possible that are each 3 ft long.
How many 3-ft long pieces are there?
How long is the left over piece of string?

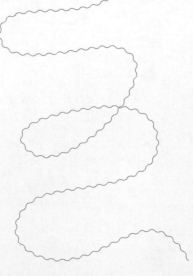

There are _____ pieces that are 3 ft long.

The left over piece of string is _____ ft long.

Basics

1 Count by fours and complete the multiplication equations.

$1 \times 4 = \boxed{}$

$2 \times 4 = \boxed{}$

$3 \times 4 = \boxed{}$

$4 \times 4 = \boxed{}$

$5 \times 4 = \boxed{}$

$6 \times 4 = \boxed{}$

$7 \times 4 = \boxed{}$

$8 \times 4 = \boxed{}$

$9 \times 4 = \boxed{}$

$10 \times 4 = \boxed{}$

2 The ones digit in the products of 4 are

_____, _____, _____, _____, or _____.

Practice

3 (a) $5 \times 4 = \boxed{}$

6×4 is $\boxed{}$ more than 5×4.

$6 \times 4 = \boxed{}$

7×4 is 8 more than $\boxed{} \times 4$.

$7 \times 4 = \boxed{}$

(b) $3 \times 4 = \boxed{}$

$4 \times 4 = \boxed{} + 4 = \boxed{}$

$8 \times 4 = 16 + \boxed{} = \boxed{}$

(c) $9 \times 4 = 40 - \boxed{} = \boxed{}$

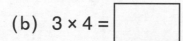

4 Each dragonfly has 4 wings.
How many wings are on 8 dragonflies?

$\boxed{} \times \boxed{} = \boxed{}$

_____ wings are on 8 dragonflies.

5 Circle products of 4.

32   **16**   **18**   **23**   **12**   **15**   **24**

Basics

1 $8 \times 4 = $ ☐

$4 \times 8 = $ ☐

2 (a) $3 \times 4 = $ ☐ $4 \times 3 = $ ☐

(b) $6 \times 4 = $ ☐ $4 \times 6 = $ ☐

(c) $5 \times 4 = $ ☐ $4 \times 5 = $ ☐

(d) $2 \times 4 = $ ☐ $4 \times 2 = $ ☐

(e) $7 \times 4 = $ ☐ $4 \times 7 = $ ☐

(f) $10 \times 4 = $ ☐ $4 \times 10 = $ ☐

(g) $9 \times 4 = $ ☐ $4 \times 9 = $ ☐

(h) $4 \times 4 = $ ☐

3 Each semi truck has 6 wheels.

How many wheels do 4 of these semi trucks have?

☐ $\times$ ☐ $=$ ☐

4 semi trucks have _____ wheels.

Practice

4 Match.

4 × 3	16	9 × 4
1 × 4	32	4 × 5
4 × 10	8	4 × 7
7 × 4	36	10 × 4
2 × 4	4	8 × 4
4 × 6	20	6 × 4
4 × 4	28	4 × 4
4 × 9	12	3 × 4
5 × 4	40	4 × 2
4 × 8	24	4 × 1

Basics

1 Alex is transplanting 32 tomato seedlings.

(a) If he puts 4 seedlings in each row,
how many rows will he have?

| | × 4 = 32 | 32 ÷ 4 = | |

He will have _____ rows.

(b) If he wants to put them equally in 4 rows,
how many seedlings should he put in each row?

4 × | | = 32 | 32 ÷ 4 = | |

He should put _____ seedlings in each row.

2 Mei is planting 4 melon seeds in each pot.

(a) If she has 6 pots, how many melon seeds does she need?

| | × | | = | |

She needs _____ seeds.

(b) If she has 24 seeds, how many pots does she need?

| | ÷ | | = | |

She needs _____ pots.

Practice

 3   ▭ × 4 = 12

12 ÷ 4 = ▭

▭ × 4 = 32

32 ÷ 4 = ▭

▭ × 4 = 4

4 ÷ 4 = ▭

▭ × 4 = 16

16 ÷ 4 = ▭

▭ × 4 = 24

24 ÷ 4 = ▭

▭ × 4 = 36

36 ÷ 4 = ▭

▭ × 4 = 8

8 ÷ 4 = ▭

▭ × 4 = 28

28 ÷ 4 = ▭

▭ × 4 = 20

20 ÷ 4 = ▭

▭ × 4 = 40

40 ÷ 4 = ▭

4 (a) ▭ ÷ 4 = 4

(b) ▭ ÷ 4 = 9

(c) ▭ ÷ 4 = 7

(d) ▭ ÷ 4 = 5

(e) ▭ ÷ 4 = 6

(f) ▭ ÷ 4 = 8

5 Jasper, Landon, Malik, and Wyatt share a box of 40 colored pencils equally. How many pencils does boy get?

40 ÷ [ ] = [ ]

Each boy gets _____ pencils.

6 There are 16 basketballs.
Grace puts them equally in 4 bags.
How many basketballs are in each bag?

[ ] ÷ [ ] = [ ]

_____ basketballs are in each bag.

7 Mary has 28 pounds of flour.
She puts 4 pounds of flour in each bag.
How many bags does she need?

[ ] ◯ [ ] 4 ◯ [ ]

She needs _____ bags.

3 Multiply or divide.

$3 \times 4 =$ ___ **I**	$4 \div 4 =$ ___ **E**	$12 \div 4 =$ ___ **F**
$12 \div 3 =$ ___ **R**	$6 \times 4 =$ ___ **T**	$20 \div 4 =$ ___ **S**
$4 \times 8 =$ ___ **E**	$24 \div 3 =$ ___ **K**	$9 \times 4 =$ ___ **C**
$18 \div 2 =$ ___ **P**	$28 \div 4 =$ ___ **D**	$18 \div 3 =$ ___ **E**
$4 \times 4 =$ ___ **A**	$8 \div 4 =$ ___ **W**	$5 \times 5 =$ ___ **H**
$3 \times 6 =$ ___ **O**	$40 \div 4 =$ ___ **A**	$45 \div 5 =$ ___ **P**
$10 \times 6 =$ ___ **E**	$3 \times 7 =$ ___ **H**	$5 \times 3 =$ ___ **A**

Joke: Why did the boy eat his math homework?

Write the letters that match the answers above to find out.

25	32	14	21	1	10	4	7	29	12	24

2	16	5	26	15	20	9	12	6	36	60

45	30	18	3	27	36	16	8	32	0	50

4 Ana bought 4 packs of pens.
There were 5 pens in each pack.
How many pens did she buy?

She bought _____ pens.

5 Dexter bought 4 packs of markers.
He bought 32 markers in all.
How many markers are in each pack?

There are _____ markers in each pack.

6 Gary bought 4 pints of strawberries for $16.

(a) How much does 1 pint of strawberries cost?

1 pint of strawberries costs $_____.

(b) How much would 8 pints of strawberries cost?

8 pints of strawberries would cost $_____.

Challenge

7 Mrs. Garcia made 34 empanadas.

(a) She wants to put them equally into 4 boxes.
How many empanadas will be left over?

_____ empanadas will be left over.

(b) She gives away 2 full boxes of empanadas.
How many empanadas does she have left?

She has _____ empanadas left.

8 Clara has enough money to buy 4 comic books at $6 each.
Instead, she uses the same amount of money to buy some notebooks that cost $3 each.
How many notebooks does she buy?

She buys _____ notebooks.

Exercise 9

Check

1 Find the values.

(a) $37 + 5 = \boxed{}$

(b) $659 + 7 = \boxed{}$

(c) $250 + 80 = \boxed{}$

(d) $598 + 50 = \boxed{}$

(e) $65 + 98 = \boxed{}$

(f) $99 + 52 = \boxed{}$

(g) $62 - 7 = \boxed{}$

(h) $691 - 8 = \boxed{}$

(i) $640 - 60 = \boxed{}$

(j) $351 - 80 = \boxed{}$

(k) $100 - 32 = \boxed{}$

(l) $303 - 97 = \boxed{}$

2 Write the missing numbers.

(a) $\boxed{} \div 3 = 7$

(b) $24 \div \boxed{} = 4$

(c) $5 \times \boxed{} = 40$

(d) $\boxed{} \times 2 = 16$

(e) $5 = 25 \div \boxed{}$

(f) $12 = 4 \times \boxed{}$

3 Olga is planting corn, beans, and squash together.

(a) She is putting 5 corn seeds in each mound of dirt.
There are 8 mounds.
How many corn seeds does she use?

(b) When the corn is 5 ft tall, she plants the same number of bean seeds around each mound.
She uses 32 bean seeds.
How many bean seeds does she plant around each mound?

(c) Then, she plants three yellow squash seeds and three green squash seeds around each mound.
How many of each type of squash seed is around all 8 mounds?

(d) How many squash seeds did she plant in all?

4 This table lists the cost of some bags of fruit.

Apples	$2
Grapes	$5
Bananas	$4
Oranges	$3
Lychee	$10

(a) How much do 6 bags of grapes cost?

_____

(b) How much do 6 bags of apples cost?

_____

(c) Melissa buys 6 bags of grapes and 6 bags of apples. How much does she spend?

_____

(d) How many bags of lychees can Travis buy with $40?

_____

(e) How many bags of grapes can Travis buy with $40?

_____

(f) How many more bags of grapes can Travis buy than bags of lychees?

_____

(g) What is the total cost of 2 bags of each kind of fruit?

5 Complete the cross-number puzzle.

32	÷		=	8
−		+		+
	÷		=	10
=		=		=
	+		=	18

6 ◆ + ● + ✴ = 16

● + ◆ + ◆ = 18

● + ● = 8

45 ÷ ✴ = ☐

7 How many different products can be made
by multiplying any two numbers below?

2   3
  4   5

Chapter 10 Money

Basics

1 1 dollar = 100 cents

$1 = ☐ ¢

2 (a) Write the total cents for each set of coins.

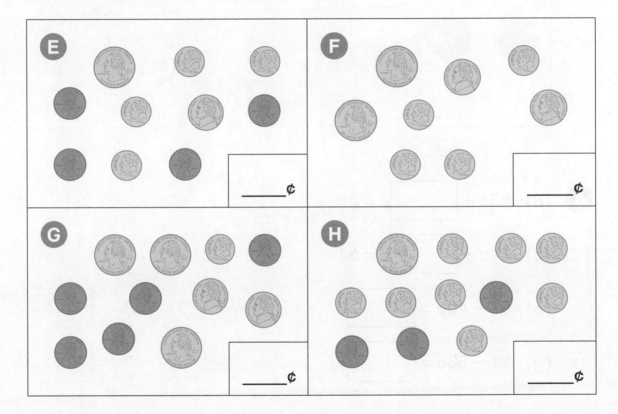

E _____¢

F _____¢

G _____¢

H _____¢

(b) Which sets have the same amount of money as 1 dollar?

(c) Which set has more than 1 dollar?

(d) What coins can you add to the set that has less than 1 dollar to make $1?

Practice

3 Circle to separate the coins into 4 sets of $1.

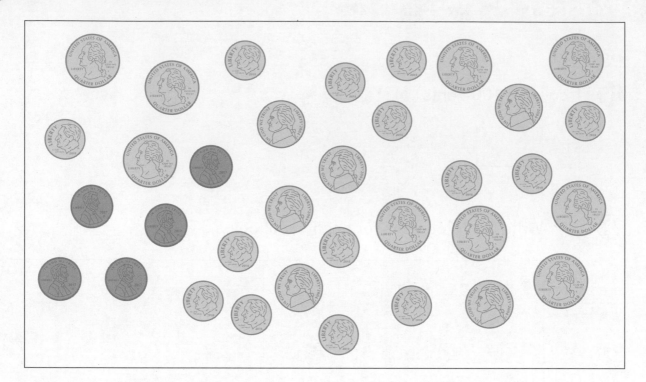

4 (a) 85¢ + [   ]¢ = 100¢   (b) 93¢ + [   ]¢ = $1

(c) 24¢ + [   ]¢ = $1   (d) 7¢ + [   ]¢ = $1

(e) $1 − 40¢ = [   ]¢   (f) $1 − 25¢ = [   ]¢

(g) $1 − 53¢ = [   ]¢   (h) $1 − 22¢ = [   ]¢

5 Santino pays for a snack that cost 62¢ with $1.
How much change does he get?

Basics

 Match.

$2.45

$0.65

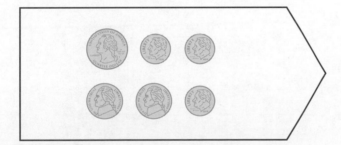

$2.56

$6.07

$3.40

2 Match.

35 cents	$8.05
35 dollars	$12.21
8 dollars and 50 cents	$35.00
12 dollars and 21 cents	$35.02
8 dollars and 5 cents	$0.35
35 dollars and 2 cents	$8.50

Practice

3 Write the amounts of money.

Forty-five cents	$_____.____
One dollar and ninety cents	$_____.____
Twenty-three dollars and eighteen cents	$_____.____
Ninety dollars and forty-seven cents	$_____.____
Six dollars and six cents	$_____.____
Nine cents	$_____.____
Forty dollars and four cents	$_____.____
Thirty-eight dollars	$_____.____

4 Write the amount of money in each set.

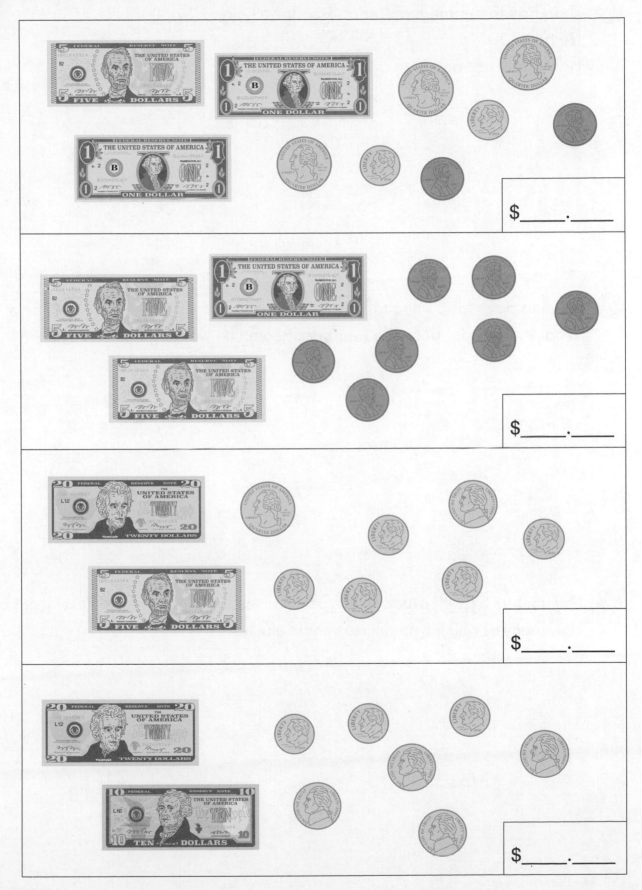

$\$$_____._____

$\$$_____._____

$\$$_____._____

$\$$_____._____

Challenge

5 Bron has the same number of 5-dollar bills and 1-dollar bills.
He has $30.
How many of each type of bill does he have?

6 Santino pays for a snack that cost 62¢ with $1.
What is the least possible number of coins he would get in change?

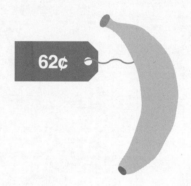

62¢

7 Mayam has $1 with exactly two types of coins (quarters, dimes, nickels, or pennies).
How many of which type of coins could she have?

Basics

1 (a) 100¢ = $1.00

(b) 200¢ = $____.____

25¢ = $0.25

5¢ = $____.____

125¢ = $____.____

205¢ = $____.____

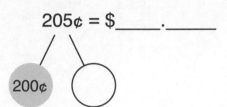

2 Match.

735¢	$6.05
605¢	$7.35
65¢	$6.50
650¢	$0.05
700¢	$7.65
5¢	$0.65
765¢	$7

Practice

3 Write the amount of money.

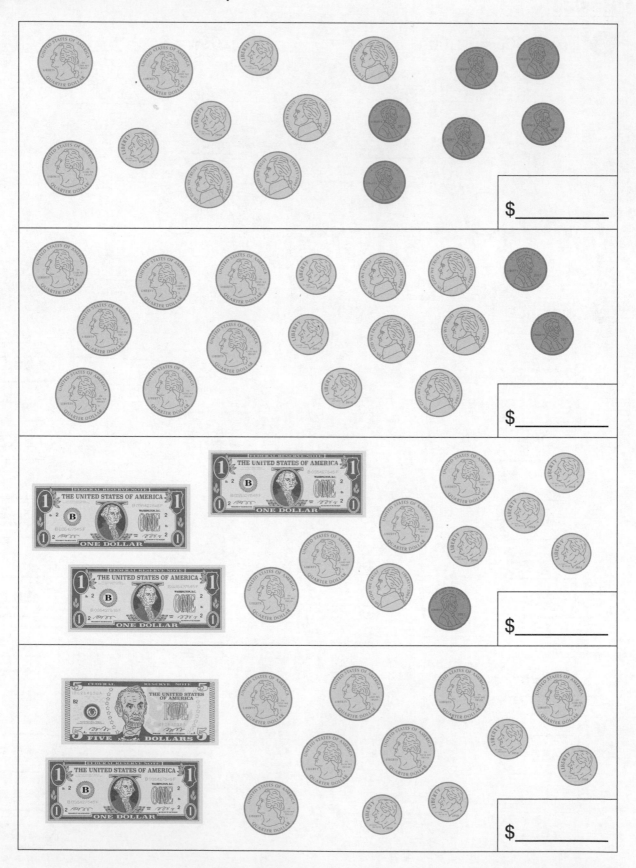

$_____

$_____

$_____

$_____

4 Complete the tables.

Dollars	Cents
$4.32	432¢
$0.24	
$0.08	
$3.40	
$8.02	
$9.56	
$6.20	
$8	

Cents	Dollars
231¢	$2.31
870¢	
708¢	
428¢	
75¢	
40¢	
400¢	
8¢	

5 (a) _____ quarters make $2.00

(b) _____ dimes make $2.00

(c) _____ nickels make $2.00

(d) _____ pennies make $2.00

(e) _____ quarters make $2.75

(f) _____ dimes make $3.50

(g) _____ nickels make $1.60

6 Write the number of each type of bill or coin to make the given amount.
Show three different ways for each amount.

Amount	$5	$1	25¢	10¢	5¢
$1.30		1	1		1
$2.75					
$7.45					
$10					

Challenge

7 Kiara saves 1 quarter the first month, 2 quarters the second
month, 3 quarters the third month, and so on.
How much money will she save in 5 months?

Basics

1 (a) Complete the tables and write >, <, or = in each ◯.

Amount	Dollars	Cents
$9.45	9	
$7.62		62

$9 ◯ $7

$9.45 ◯ $7.62

Amount	Dollars	Cents
$3.45		45
$3.62	3	

$3 ◯ $3

45¢ ◯ 62¢

$3.45 ◯ $3.62

(b) Arrange the amounts of money in order from least to greatest.

$9.45	$7.62	$3.45	$3.62

Practice

2 (a) Which set has more money? _____

A

$_____

B

$_____

(b) Which set has less money? _____

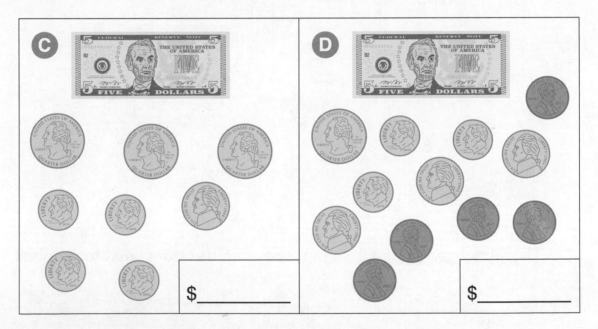

C

$_____

D

$_____

(c) Arrange the sets in order from greatest to least amount of money.

_____ _____ _____ _____

3 Circle the item that costs the most.

4 Circle the item that costs the least.

5 Write >, <, or = in each ◯.

(a) $5.62 ◯ $6.94

(b) $6.81 ◯ $6.48

(c) $7.03 ◯ $3.07

(d) $60.20 ◯ $6.65

(e) 854¢ ◯ 812¢

(f) 123¢ ◯ $123

(g) $1.02 ◯ 102¢

(h) 8¢ ◯ $8

(i) $0.71 ◯ 83¢

(j) $4 ◯ 349¢

6 The table shows how much money 5 people saved.

Arrange the amounts of money in order.

Start with the greatest amount.

Kalama	$6.23
Nolan	$5.99
Madison	$8.34
Pablo	$6.32
Jody	$9.34

7 Circle the amounts of money that are greater than $2.45 but less than $6.

$2.35   **$2.85**   **$5.62**   **$6.07**   **$3.86**   **$7.42**

Challenge

8 Riya has 1 five-dollar bill, 2 one-dollar bills, 5 quarters, and 12 dimes.

Nora has 5 one-dollar bills, 12 quarters, and 5 dimes.

Who has more money?

Check

1 Practice addition and subtraction.

198 + 49 = ☐

327 + 327 = ☐

483 + 217 = ☐

304 − 38 = ☐

932 − 762 = ☐

425 − 89 = ☐

2 Color coins in each set to make $1 two different ways.

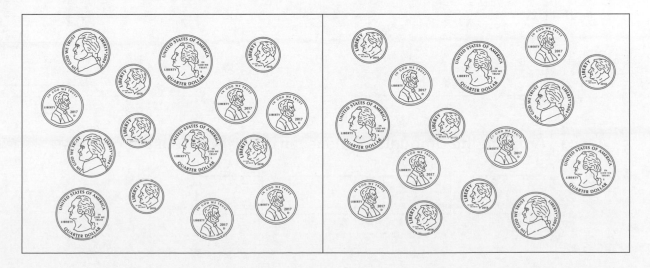

3 (a) Write the amount of money in each set.

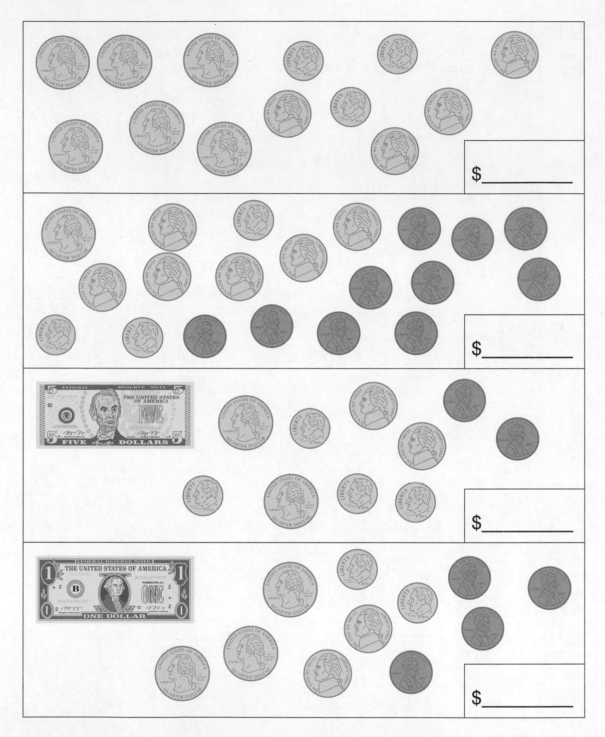

$ _____

$ _____

$ _____

$ _____

(b) Arrange the amounts of money in order from least to greatest.

4 Write the number of each type of bill or coin to make the given amount. Show four different ways for each amount.

Amount	$5	$1	25¢	10¢	5¢
$1.80		1	3		1
$3.35					
$5					
$8.50					

5 Write >, <, or = in each ◯.

(a) $5.62 ◯ $6.94

(b) $0.10 ◯ 1¢

(c) $3.04 ◯ $4.90

(d) $26 ◯ 260¢

(e) 824¢ ◯ $842

(f) 430¢ ◯ $4.30

Challenge

6 Natalia bought a toy for $0.85.
She paid with a 5-dollar bill.
She got 3 bills and 8 coins in change.
What coins were they?

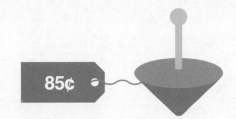

7 Find 2 ways to make $1 using 19 coins
(quarters, dimes, nickels, or pennies) exactly.

Basics

1 (a) Add $5.70 and $2.25.

$5.70 $\xrightarrow{+\ \$2}$ \$ ☐ $\xrightarrow{+\ 25¢}$ \$ ☐

(b) Add $5.70 and $2.30.

$5.70 $\xrightarrow{+\ \$2}$ \$ ☐ $\xrightarrow{+\ 30¢}$ \$ ☐

(c) Add $5.70 and $2.45.

$5.70 $\xrightarrow{+\ \$2}$ \$ ☐ $\xrightarrow{+\ 45¢}$ \$ ☐

2 65¢ + 35¢ = $ ☐

65¢ + 50¢ = $ ☐

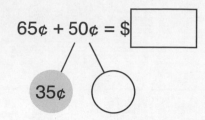

35¢   ◯

$4.65 + $3 = $ ☐

$4.65 + $0.35 = $ ☐

$4.65 + $0.50 = $ ☐

35¢   ◯

$4.65 + $3.50 = $ ☐

Practice

3 (a) $0.85 + $3 = $ ☐   (b) $3.45 + $2 = $ ☐

(c) $6.75 + 15¢ = $ ☐   (d) $2.45 + 55¢ = $ ☐

(e) $1.60 + 50¢ = $ ☐   (f) $4.85 + 40¢ = $ ☐

4 Write the missing numbers.

(a) $2.35 ──+ $4──▶ $ ☐ ──+ 65¢──▶ ☐

$2.35 + $4.65 = ☐

(b) $7.20 ──+ $2──▶ $ ☐ ──+ 95¢──▶ ☐

$7.20 + $2.95 = ☐

5 Add.

30¢ + 45¢ = ☐	$1.22 + 20¢ = ☐	$2.50 + $1.40 = ☐
$8.60 + $0.04 = ☐	$4.11 + $1.50 = ☐	$5.55 + $2.80 = ☐
75¢ + $3.51 = ☐	$7.10 + $2.90 = ☐	$1.89 + $0.75 = ☐
$1.95 + $1.95 = ☐	$1.49 + $0.99 = ☐	$9.19 + 14¢ = ☐
$3.40 + $2.25 = ☐	$0.95 + $2.60 = ☐	$5.10 + 45¢ = ☐

How many hearts does an octopus have?

Color the spaces that contain the answers to find out.

$2.20	$6.00	$2.48	$8.35	$9.33	75¢	$5.23	99¢
$9.60	$3.33	$9.99	$7.80	$1.25	$4.26	$4.76	$2.75
$2.59	19¢	$8.90	$5.98	22¢	$5.65	$7.00	$9.50
$7.25	$4.02	50¢	$1.42	$3.90	$5.55	$2.20	$1.85
$6.30	$2.45	$2.40	$3.77	$9.20	$3.90	54¢	$9.50
$8.10	57¢	$8.15	$6.12	$7.65	$10.00	$4.19	$3.89
$5.27	$1.99	$5.61	$8.64	$2.64	$3.55	$6.15	$7.45

$1.80 $3.20 $3.75 $0.85

6 Aisha bought the cinnamon roll and the pack of gum.
How much did she spend?

7 Carter bought the muffin and the pretzel.
How much did he spend?

8 Dexter bought 2 of the items above and paid with a five-dollar bill.
He did not get any change.
Which two items did he buy?

9 Valentina bought all 4 items above.
How much did she spend?

Basics

1 (a) Subtract $1.20 from $5.50.

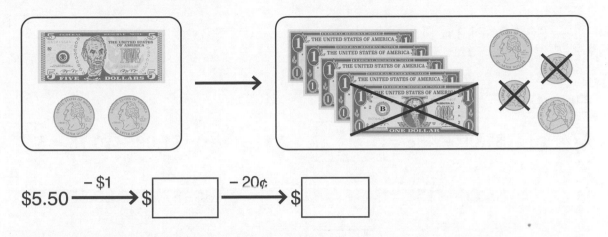

$5.50 ——$-\$1$——▶ $ [ ] ——$-20¢$——▶ $ [ ]

(b) Subtract $1.50 from $5.50.

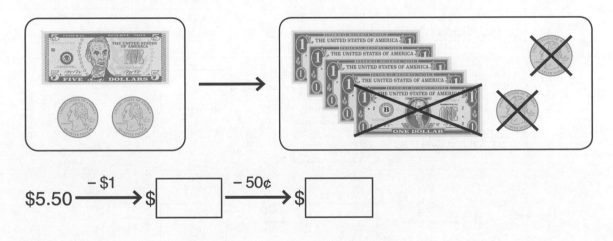

$5.50 ——$-\$1$——▶ $ [ ] ——$-50¢$——▶ $ [ ]

(c) Subtract $1.75 from $5.50.

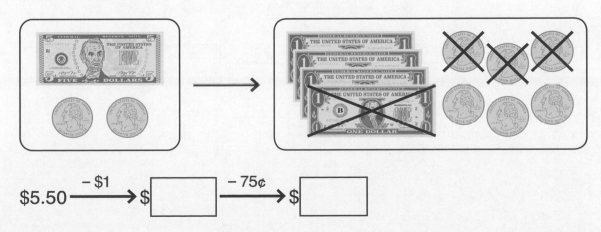

$5.50 ——$-\$1$——▶ $ [ ] ——$-75¢$——▶ $ [ ]

2 $6.00 − 35¢ = $ [ ] $6.20 − 35¢ = $ [ ]

( ) ── 100¢   ( ) ── 100¢

$6.20 − $3 = $ [ ] $6.20 − $3.35 = $ [ ]

Practice

3 (a) $1.00 − 45¢ = [ ] ¢   (b) $1.00 − $0.75 = $ [ ]

(c) $5.00 − 15¢ = $ [ ]   (d) $7.00 − $0.55 = $ [ ]

(e) $1.10 − 25¢ = $ [ ]   (f) $4.15 − $0.40 = $ [ ]

4 Write the missing numbers.

(a) $7.00 ──− $4──→ $ [ ] ──− 65¢──→ $ [ ]

$7.00 − $4.65 = $ [ ]

(b) $7.95 ──− $2──→ $ [ ] ──− 20¢──→ $ [ ]

$7.95 − $2.20 = $ [ ]

(c) $7.20 ──− $2──→ $ [ ] ──− 95¢──→ $ [ ]

$7.20 − $2.95 = $ [ ]

5 Subtract.

45¢ − 20¢ = ☐ **E**	$7.50 − $6.40 = ☐ **C**	$4.85 − $2.50 = ☐ **H**
$8.75 − $3.60 = ☐ **O**	$1.25 − 80¢ = ☐ **B**	$9.00 − $5.30 = ☐ **E**
$3.99 − $1.99 = ☐ **Y**	$8.70 − $4.60 = ☐ **G**	$5.10 − 90¢ = ☐ **A**
$2.25 − $1.75 = ☐ **U**	$9.65 − $1.70 = ☐ **S**	$6.22 − $0.99 = ☐ **L**
$1.20 − 27¢ = ☐ **H**	$5.87 − $2.36 = ☐ **B**	$4.89 − $3.80 = ☐ **T**

Joke: Why isn't it fun to play basketball with pigs?

Write the letters that match the answers above to find out.

45¢	25¢	$1.10	$4.20	50¢	$7.95	$3.70	$5.22	$1.09	$2.35	$3.70	$2.00

93¢	$5.15	$4.10	$1.01	$1.09	$2.35	25¢	99¢	$3.51	$4.20	$5.23	$5.23

$1.80 $3.20 $3.75 $0.85 SPEARMINT

6 Austin bought the soft pretzel.
He paid with a five-dollar bill.
How much change did he receive?

7 How much more is the cinnamon roll than the pack of gum?

Challenge

8 Papina bought 2 items above and paid with a five-dollar bill.
She got 40¢ in change.
Which two items did she buy?

Check

1 Color bills and coins to make the amounts of money.

2 Write the amounts of money.

12 dimes and 3 nickels	$_____
2 one-dollar bills and 5 quarters	$_____
1 five-dollar bill, 2 one-dollar bills, and 3 quarters	$_____
7 quarters and 7 dimes	$_____
25 nickels and 25 pennies	$_____

3 The answer to each problem is the first number in the next problem.
Go through all the calculations in order to reach the goal of $10.00.

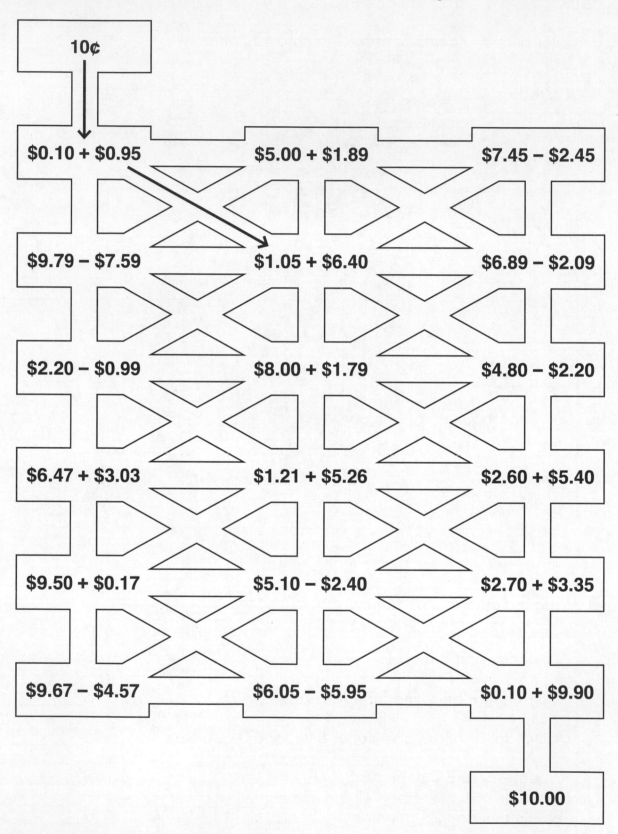

10¢

$0.10 + $0.95 $5.00 + $1.89 $7.45 − $2.45

$9.79 − $7.59 $1.05 + $6.40 $6.89 − $2.09

$2.20 − $0.99 $8.00 + $1.79 $4.80 − $2.20

$6.47 + $3.03 $1.21 + $5.26 $2.60 + $5.40

$9.50 + $0.17 $5.10 − $2.40 $2.70 + $3.35

$9.67 − $4.57 $6.05 − $5.95 $0.10 + $9.90

$10.00

4 A toy car costs $6.20.

A toy robot costs $1.55 more than the toy car.

How much does the toy robot cost?

5 Andrei has $6.10.

He has 80¢ more than Daren.

How much money does Daren have?

Challenge

6 Nina bought a hair bow for $0.80, a bracelet for $0.75, and a purse for $2.

She had $4.30 left.

How much money did she have at first?

Challenge

7 Circle to separate the bills and coins into 4 sets of $3.00.

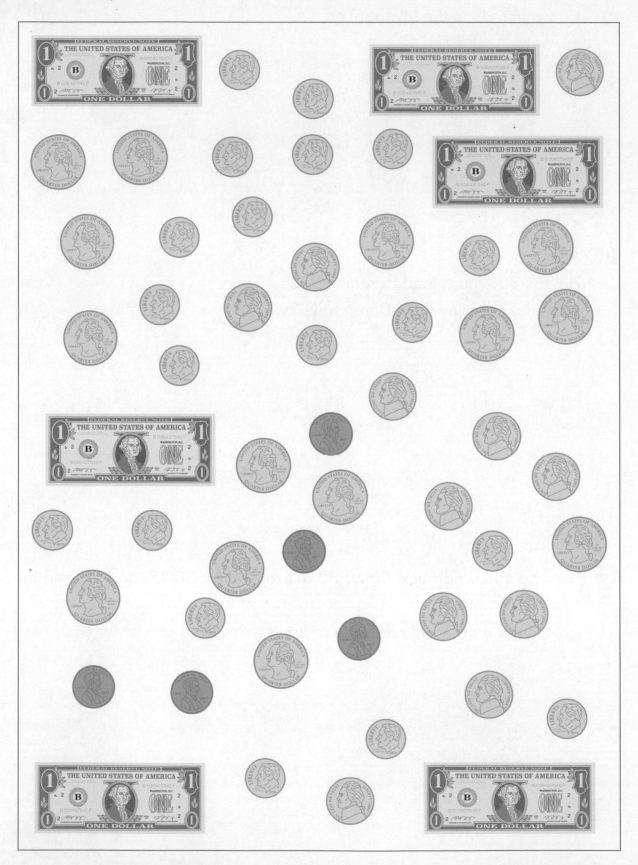

Chapter 11 Fractions

Basics

1 Draw a line to divide each figure into 2 equal parts.

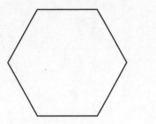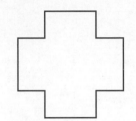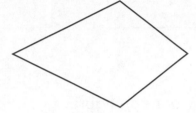

2 Draw lines to divide each figure into 4 equal parts.

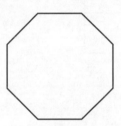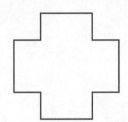

3 Check the figures that are $\frac{1}{2}$ shaded.

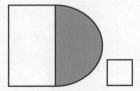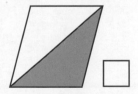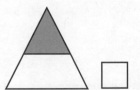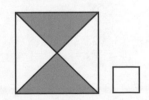

4 Check the figures that are $\frac{1}{4}$ shaded.

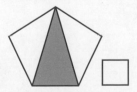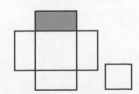

Practice

5 Color $\frac{1}{2}$ of each figure.

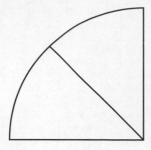 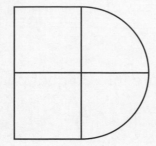 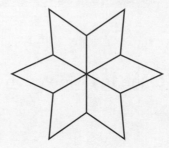

6 Color $\frac{1}{4}$ of each figure.

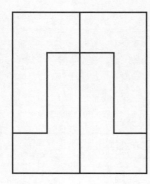 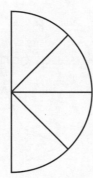 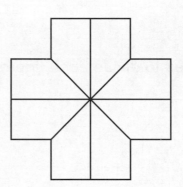

Challenge

7 Draw more squares to make a figure that is $\frac{1}{4}$ shaded.

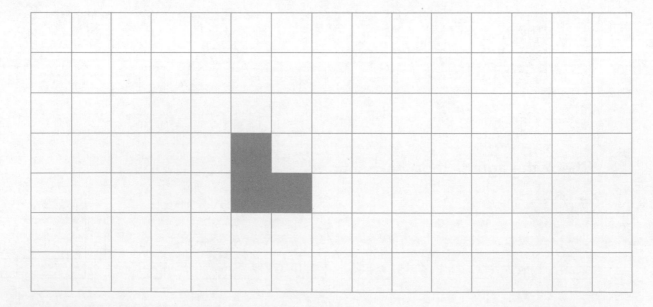

Exercise 2

Basics

 (a)

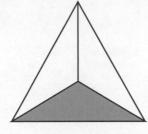

This triangle is divided into 3 equal parts.

$\frac{1}{3}$ of the triangle is shaded.

One-third means _____ out of _____ equal parts.

(b)

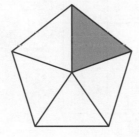

This figure is divided into _____ equal parts.

Each part is one-_____ of the whole.

$\boxed{\dfrac{1}{}}$ of the figure is shaded.

(c)

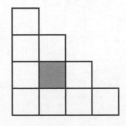

This figure is divided into _____ equal parts.

Each part is one-_____ of the whole.

$\boxed{\dfrac{}{10}}$ of the figure is shaded.

(d)

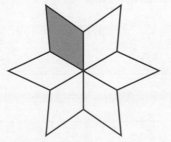

This figure is divided into _____ equal parts.

$\boxed{\dfrac{}{}}$ of the figure is shaded.

One-_____ means 1 out of 6 equal parts.

Practice

2 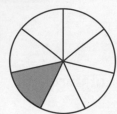 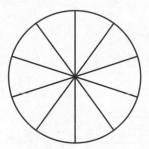 of the circle is shaded.

One-_____ of the circle is shaded.

3 Color to show the given fraction.

$\frac{1}{3}$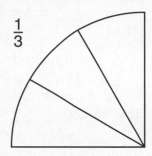

$\frac{1}{9}$

$\frac{1}{8}$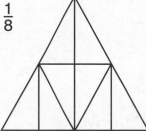

$\frac{1}{10}$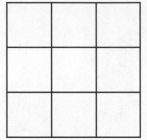

4 Write the fraction in words.

$\frac{1}{3}$	
$\frac{1}{5}$	
$\frac{1}{8}$	
$\frac{1}{10}$	
$\frac{1}{12}$	

5 Match the circles to the fraction shaded.

  $\dfrac{1}{6}$

 $\dfrac{1}{9}$

 $\dfrac{1}{12}$

 $\dfrac{1}{10}$

 $\dfrac{1}{7}$

 $\dfrac{1}{4}$

 $\dfrac{1}{3}$

 $\dfrac{1}{8}$

 $\dfrac{1}{5}$

 $\dfrac{1}{2}$

$\dfrac{1}{11}$

Challenge

6 Draw lines and shade one part to show the given fractions.
The first one is done for you.

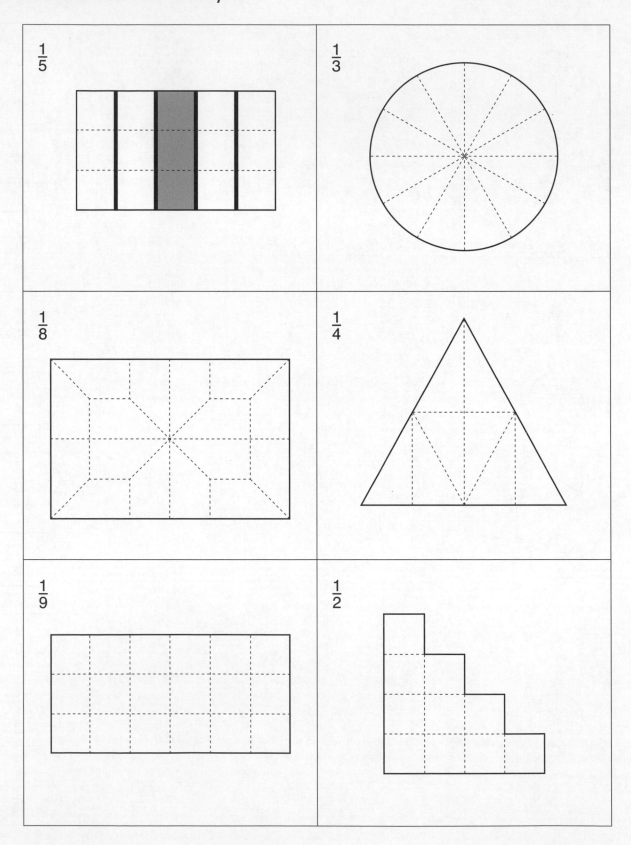

$\frac{1}{5}$

$\frac{1}{3}$

$\frac{1}{8}$

$\frac{1}{4}$

$\frac{1}{9}$

$\frac{1}{2}$

Basics

1 (a)

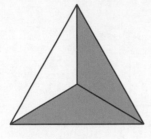

The triangle is divided into 3 equal parts.

2 out of _____ parts are shaded.

$\frac{2}{3}$ of the triangle is shaded.

(b)

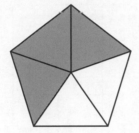

$\frac{3}{5}$ of the figure is shaded.

$\frac{3}{5}$ is _____ out of _____ equal parts.

There are _____ one-fifths in $\frac{3}{5}$.

(c)

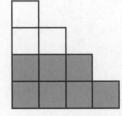

_____ out of 10 parts are shaded.

$\boxed{\dfrac{\quad}{10}}$ of the figure is shaded.

There are 7 one-tenths in $\boxed{\dfrac{\quad}{\quad}}$.

(d)

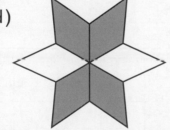

_____ out of 6 parts are shaded.

$\boxed{\dfrac{\quad}{\quad}}$ of the figure is shaded.

There are _____ one-sixths in $\boxed{\dfrac{4}{\quad}}$.

Practice

2 (a) There are _____ one-sevenths in $\frac{5}{7}$.

(b) There are _____ one-twelfths in 1.

(c) Write the fraction for seven elevenths. ▢

3 Color to show the given fraction.

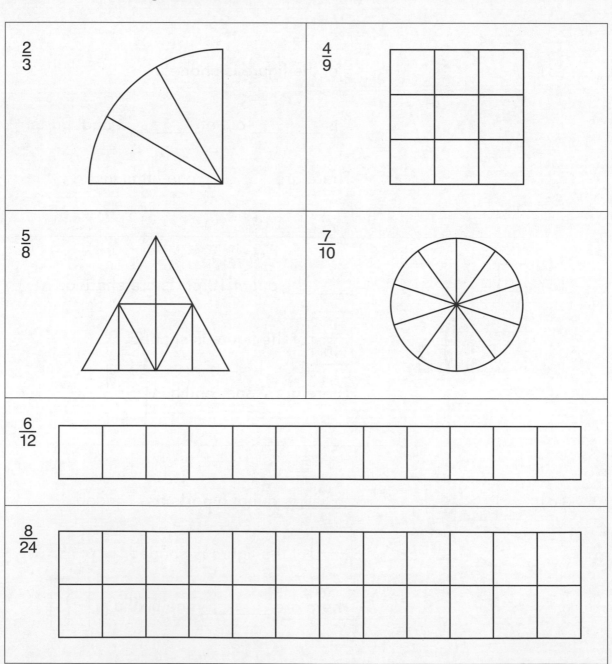

4 What fraction of each figure is shaded?

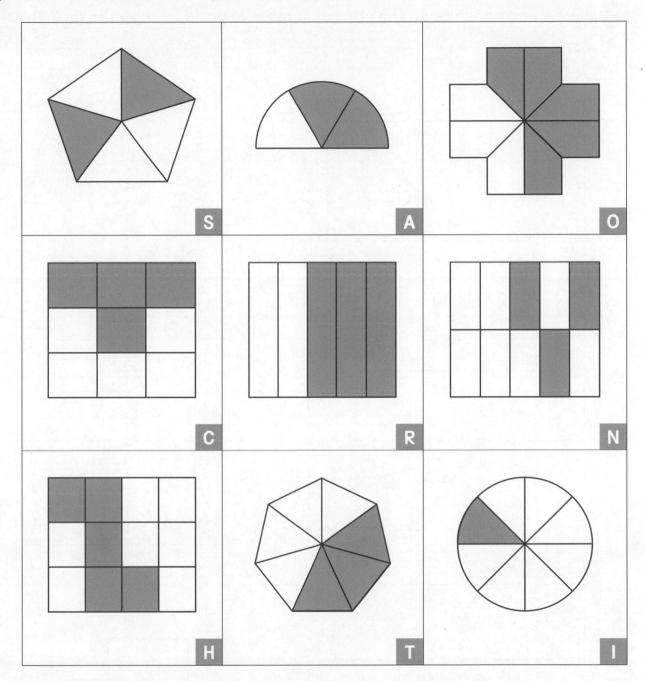

What animal's eye is bigger than its brain?

Write the letters that match the answers above to find out.

$\frac{2}{3}$	$\frac{3}{10}$	$\frac{6}{9}$	$\frac{5}{8}$	$\frac{2}{5}$	$\frac{3}{7}$	$\frac{3}{5}$	$\frac{1}{8}$	$\frac{4}{9}$	$\frac{5}{12}$

Challenge

5 Draw lines and shade the correct number of parts to show the given fractions.

$\dfrac{2}{3}$

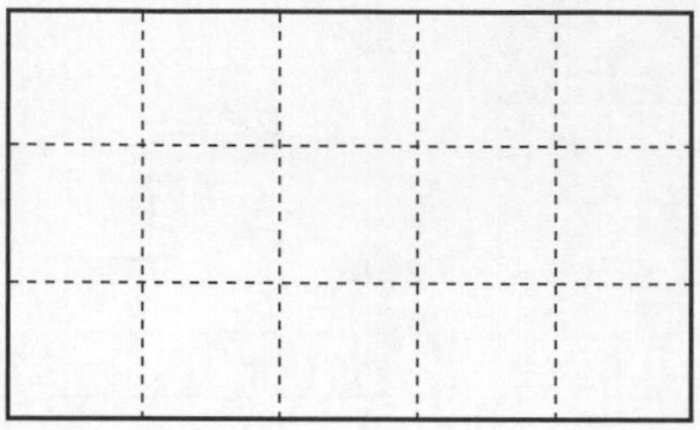

$\dfrac{5}{6}$

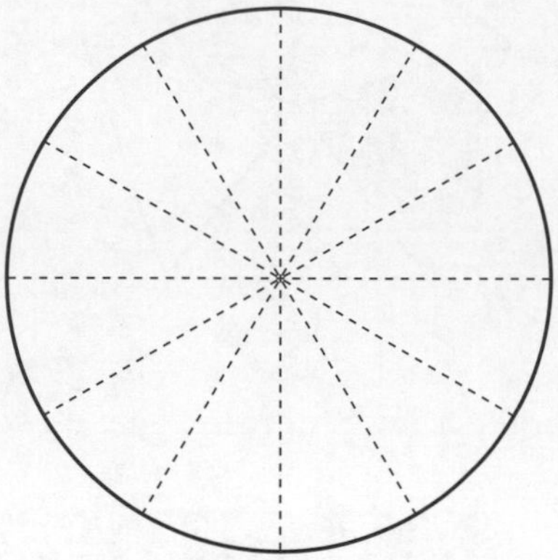

$\dfrac{5}{8}$

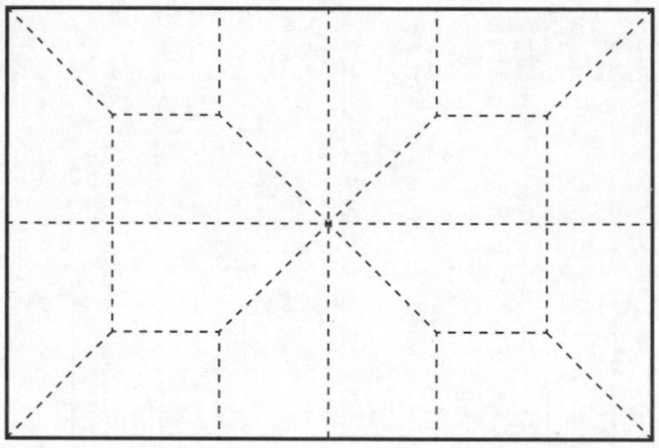

$\dfrac{3}{4}$

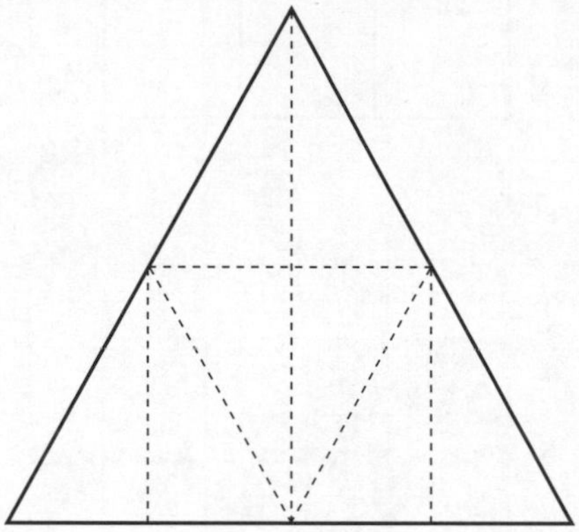

$\dfrac{4}{9}$

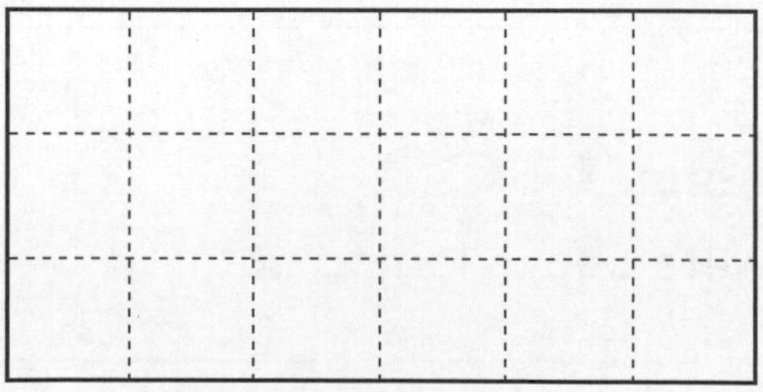

$\dfrac{3}{5}$

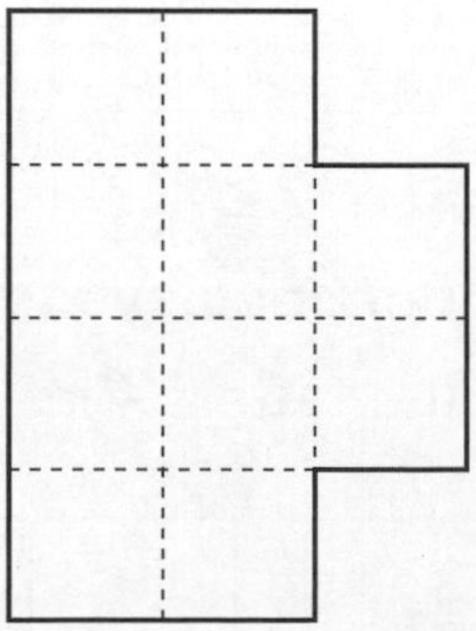

Basics

1

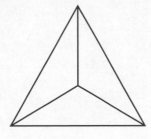

Color $\frac{1}{3}$ of the triangle red.

Color the rest of the triangle blue.

☐ of the triangle is colored blue.

$\frac{1}{3}$ and ☐ make 1 whole triangle.

2

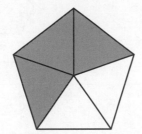

$\frac{3}{5}$ of the figure is shaded.

What fraction still needs to be
shaded to shade the whole figure? ☐

$\frac{3}{5}$ and ☐ make 1 whole figure.

3 (a)

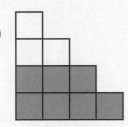

$\boxed{\dfrac{}{10}}$ and $\boxed{\dfrac{3}{}}$ make 1 whole figure.

(b)

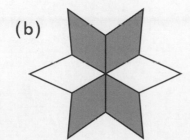

$\boxed{\dfrac{2}{}}$ and $\boxed{\dfrac{}{6}}$ make 1 whole figure.

Practice

4 Match fractions that make 1.

$\frac{6}{10}$		$\frac{3}{6}$
$\frac{7}{9}$		$\frac{2}{9}$
$\frac{3}{4}$		$\frac{7}{12}$
$\frac{3}{6}$		$\frac{4}{10}$
$\frac{5}{8}$		$\frac{5}{7}$
$\frac{5}{12}$		$\frac{1}{4}$
$\frac{2}{7}$		$\frac{3}{8}$

Challenge

5 (a) $\frac{2}{8}$, $\frac{3}{8}$, and $\boxed{}$ make 1.

(b) $\frac{4}{10}$, $\boxed{}$, and $\frac{2}{10}$ and make 1.

(c) $\boxed{}$, $\frac{4}{16}$, $\frac{4}{16}$, and $\frac{2}{16}$ make 1.

Basics

1 (a)
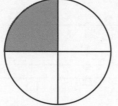   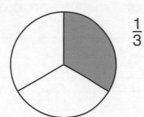
$\frac{1}{4}$ $\frac{1}{3}$

$\boxed{}$ of the circle is larger than $\boxed{}$ of the same size circle.

(b)

$\boxed{}$ of the rectangle is larger than $\boxed{}$ of the same size rectangle.

2 (a)
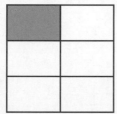 $\frac{1}{6}$ 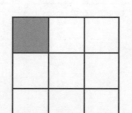 $\frac{1}{9}$

$\boxed{}$ of the square is smaller than $\boxed{}$ of the same size rectangle.

(b)
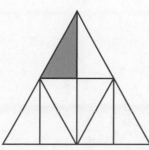

$\boxed{}$ of the triangle is smaller than $\boxed{}$ of the same-size triangle.

Practice

3 Color one part of each rectangle to show the given fraction.
Then use the rectangles to answer the problems below.

$\frac{1}{3}$

$\frac{1}{5}$

$\frac{1}{6}$

$\frac{1}{8}$

$\frac{1}{9}$

$\frac{1}{10}$

$\frac{1}{12}$

4 Circle the largest fraction.

(a) $\frac{1}{8}$ $\frac{1}{3}$ $\frac{1}{10}$

(b) $\frac{1}{6}$ $\frac{1}{8}$ $\frac{1}{9}$

(c) $\frac{1}{5}$ $\frac{1}{12}$ $\frac{1}{9}$

(d) $\frac{1}{10}$ $\frac{1}{6}$ $\frac{1}{3}$

5 Circle the smallest fraction.

(a) $\frac{1}{8}$ $\frac{1}{6}$ $\frac{1}{12}$

(b) $\frac{1}{6}$ $\frac{1}{3}$ $\frac{1}{9}$

(c) $\frac{1}{5}$ $\frac{1}{8}$ $\frac{1}{9}$

(d) $\frac{1}{10}$ $\frac{1}{6}$ $\frac{1}{5}$

6 Write the fractions in order, beginning with the smallest.

$\frac{1}{6}$ $\frac{1}{9}$ $\frac{1}{7}$ $\frac{1}{2}$ $\frac{1}{10}$

7 Write the fractions in order, beginning with the largest.

$\frac{1}{4}$ $\frac{1}{12}$ $\frac{1}{5}$ $\frac{1}{8}$ $\frac{1}{11}$

8 Darryl ate $\frac{1}{6}$ of a pizza.
His brother ate $\frac{1}{3}$ of the same pizza.
Who ate less?

9 Fang painted $\frac{1}{8}$ of a room, Debra painted $\frac{1}{5}$ of the room, and Alice painted $\frac{1}{3}$ of the room.
Who painted the most?

10 Wainani has finished reading about a third of her book.
Has she finished more or less than half of her book?

Challenge

11 Use different colors to show each fraction.
The colored parts should not overlap.
Circle the largest fraction for each.

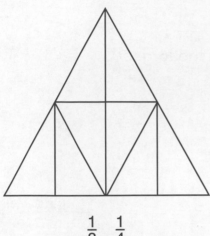

$\frac{1}{8}$ $\frac{1}{4}$

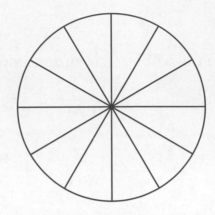

$\frac{1}{4}$ $\frac{1}{3}$ $\frac{1}{6}$

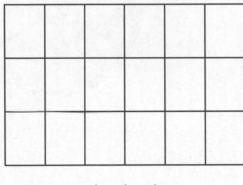

$\frac{1}{6}$ $\frac{1}{18}$ $\frac{1}{9}$

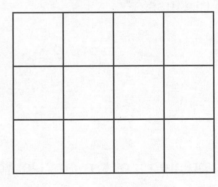

$\frac{1}{3}$ $\frac{1}{4}$ $\frac{1}{6}$ $\frac{1}{12}$

12 Is two-fourths of a shape larger than two-fifths of that same shape?

13 Circle the smaller fraction.

(a) $\frac{2}{8}$ $\frac{2}{4}$

(b) $\frac{2}{6}$ $\frac{2}{3}$

Check

1 Color to show the given fraction.

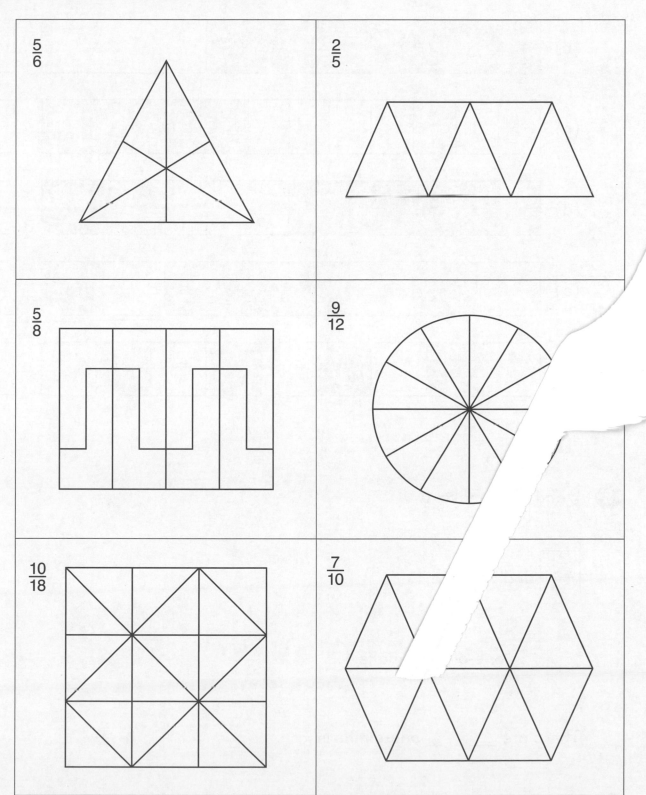

$\frac{5}{6}$

$\frac{2}{5}$

$\frac{5}{8}$

$\frac{9}{12}$

$\frac{10}{18}$

$\frac{7}{10}$

2 What fraction of each bar is shaded?

(a) ____

(b) ____

(c) ____

(d) ____

(e) ____

(f) ____

3 $\frac{3}{5}$ and ⬜ make 1.

⬜ and $\frac{5}{10}$ make 1.

_____ one-eighths make 1.

There are _____ one-ninths in $\frac{4}{9}$.

4 Circle the smallest fraction.

(a) $\frac{1}{5}$ $\frac{1}{6}$ $\frac{1}{7}$

(b) $\frac{1}{9}$ $\frac{1}{8}$ $\frac{1}{2}$

5 Write the fractions in order, beginning with the smallest.

$\frac{1}{3}$ $\frac{1}{6}$ $\frac{1}{4}$ $\frac{1}{9}$

6 Karen cut a pie into 8 equal pieces.
Her family ate 3 pieces.

(a) What fraction of the pie did her family eat?

(b) What fraction of the pie is left?

7 Eli read about one-fifth of a book on Tuesday.
He read about one-eighth of the book on Thursday.
On which day did he read more of the book?

Challenge

8 $\frac{1}{9}$ of the beads in a box are blue, $\frac{2}{9}$ of the beads are red, $\frac{3}{9}$ of the beads are green, and the rest are white.

What fraction of the beads are white?

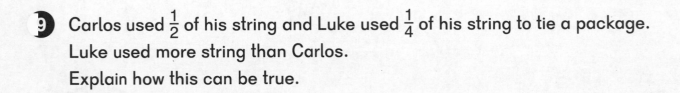

9 Carlos used $\frac{1}{2}$ of his string and Luke used $\frac{1}{4}$ of his string to tie a package.

Luke used more string than Carlos.

Explain how this can be true.

10 The figure below was made by overlapping 3 squares of the same size.

What fraction of the figure is shaded?

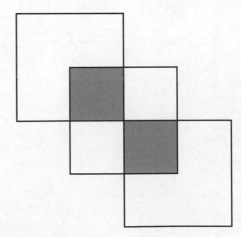

Check

1 (a) 7 hundreds, 6 tens, and 5 ones make _____.

(b) 4 hundreds, 5 ones, and 0 tens make _____.

(c) 20 less than 3 hundreds 1 ten and 8 ones is _____.

(d) _____ is 5 tens less than 3 hundreds 1 ten.

2 (a) $354 + 80 = \boxed{}$

(b) $432 - 97 = \boxed{}$

(c) $64 + \boxed{} = 100$

(d) $600 + 4 + 50 = \boxed{} - 70$

(e) $4 \times 3 = \boxed{} \div 2$

3 Write >, <, or = in the $\bigcirc$.

(a) $500 + 120 + 63 \bigcirc 80 + 503 + 200$

(b) $358 + 98 \bigcirc 453 - 97$

(c) $60 + 70 + 80 + 90 \bigcirc 120 + 50 + 40 + 80$

(d) $5 \times 2 \bigcirc 3 \times 4$

4 Find the missing numbers.

$148 + 387 = \boxed{}$ **Y**	$646 - 392 = \boxed{}$	$730 - \boxed{} = 490$ **I**
$\boxed{} + 87 = 332$ **K**	$\boxed{} - 432 = 108$ **L**	$716 - \boxed{} = 333$ **A**
$\boxed{} = 87 + 498$ **M**	$555 + \boxed{} = 872$ **Y**	$\boxed{} = 502 - 132$ **W**

Joke: What is the astronaut's favorite candy bar?
Write the letters (or space) that match the missing numbers above to find out.

585	240	540	245	535	254	370	383	317

5 Find the missing numbers.

$8 \times 3 = $ ☐ **Y**	$5 \times 6 = $ ☐ **D**	$8 \div$ ☐ $= 4$ **E**
$4 \times 7 = $ ☐ **A**	$3 \div 3 = $ ☐ **O**	$45 \div 5 = $ ☐ **I**
$2 \times 7 = $ ☐ **T**	☐ $\div 5 = 8$ **T**	$9 \div$ ☐ $= 3$ **U**
$3 \times$ ☐ $= 21$ **E**	$3 \times$ ☐ $= 15$ **C**	$24 \div 4 = $ ☐ **H**
$32 \div 4 = $ ☐ **R**	$4 \times 9 = $ ☐ **H**	$3 \times 4 = $ ☐ **I**
$5 \times$ ☐ $= 20$ **I**	☐ $\div 10 = 8$ **H**	$40 \div 4 = $ ☐ **L**
$3 \times 5 = $ ☐ **S**	☐ $= 9 \times 2$ **N**	$5 \times 7 = $ ☐ **G**

Joke: How do you know if there's an elephant under your bed?

Write the letters that match the missing numbers above to find out.

20	24	1	3	8	19	6	2	28	30	16

22	36	4	40	15	50	14	80	7	11	13

22	5	2	12	10	9	18	35	42	33	23

6

How much longer is the pencil than the crayon?

7 Fill in the blanks with cm, m, g, or kg.

A pen weighs about 15 _____.

A building has a length of about 15 _____.

A comb has a length of about 15 _____.

A boy weighs about 15 _____.

8 Can you lift something that weighs 10 lb?

9 Which is longer, $\frac{1}{2}$ of a foot or $\frac{1}{2}$ of a meter?

10 A bar of chocolate is divided equally into 10 pieces.
Victoria ate 3 of the pieces.
What fraction of the bar is left?

11

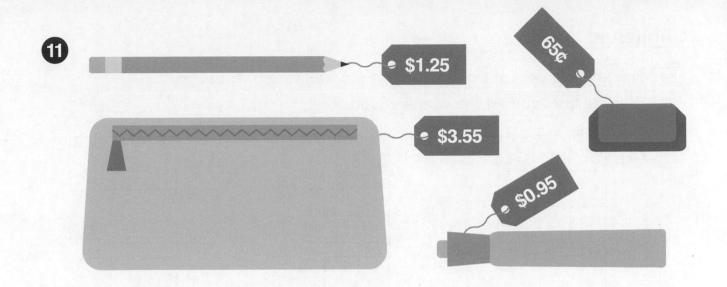

$1.25

65¢

$3.55

$0.95

(a) Lincoln has two quarters and a dime.
How much more money does he need to buy the pencil?

(b) How much more does the pencil case cost than the eraser?

(c) Rowan bought the pencil case.
She paid with a five-dollar bill.
How much change did she get?

(d) Jamal bought all 4 items.
How much did he spend?

Challenge

12 The shaded part of each square is marked with a value.
What is the value of the whole square?

(a)

(b)

(c)

(d)

(e)

(f)

13 There are 4 ropes, A, B, C, and D.
C is longer than A but shorter than B.
D is longer than B.
Match the ropes with their lengths.

Rope A

Rope B

Rope C

Rope D

20 m

20 cm

20 in

20 ft

Chapter 12 Time

Basics

1 Write the times and fill in the blanks.

It is _____ o'clock.

☐ : ☐

It is _____ minutes past 2.

It is _____ minutes to 3.

It is two forty.

☐ : ☐

It took 40 minutes for the minute hand to move from the 12 to the _____.

It is _____ minutes past 2.

Write the time in words: _____.

☐ : ☐

The hour hand is between 2 and _____.

The minute hand is between _____ and _____.

In another _____ minutes it will be 3 o'clock.

Practice

2 Cross out the clock that cannot be correct.

3 Which clock correctly shows 4:32?

☐   ☐   ☐

4 These clocks show different times in the morning.
Write the times in order from earliest time to latest time.

5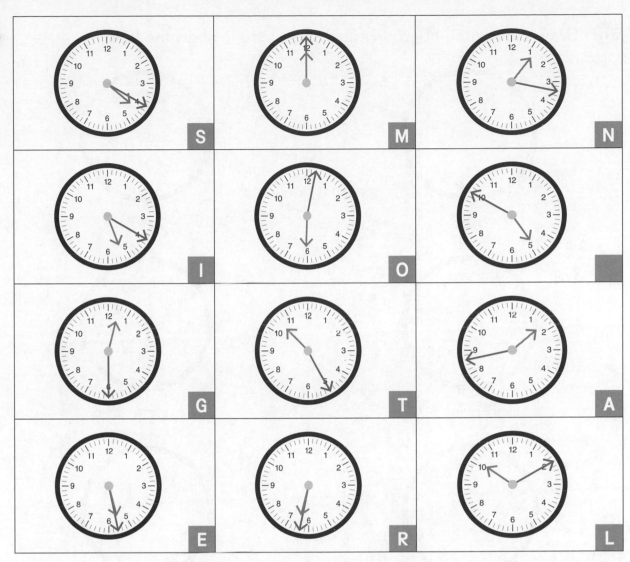

This poisonous desert-dwelling lizard from the American Southwest can live for up to 30 years.

Write the letters (or space) that match the times above to learn which animal it is.

12:30	5:20	10:10	1:43	4:50	12:00	6:02	1:17	4:20	10:25	5:28	6:32

6 Draw the minute hand on each clock face to show the time.

(a)

8:35

(b)

3:17

(c)
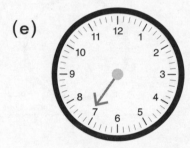
Half past 10

(d)

A quarter to 5

(e)

seven oh seven

(f)

8 minutes to 12

Challenge

7 Draw the minute hand and hour hand on each clock face to show the time.

(a)

9:45

(b)

2:30

Exercise 2

Basics

1 Write the times and fill in the blanks.

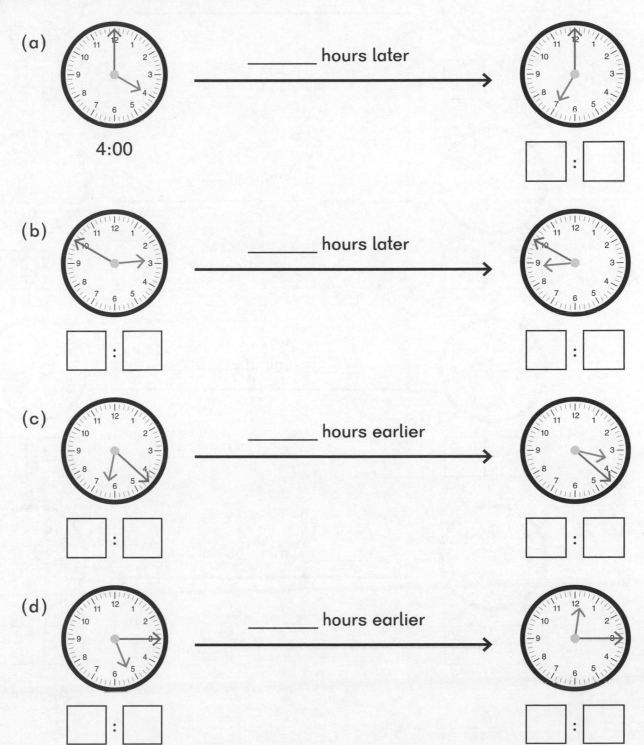

(a) _____ hours later

4:00

⬜ : ⬜

(b) _____ hours later

⬜ : ⬜

⬜ : ⬜

(c) _____ hours earlier

⬜ : ⬜

⬜ : ⬜

(d) _____ hours earlier

⬜ : ⬜

⬜ : ⬜

2 Write the times and fill in the blanks.

(a)

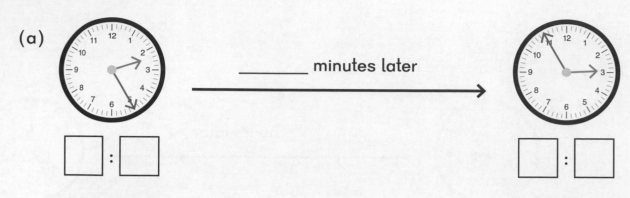

_____ minutes later

☐ : ☐   ☐ : ☐

(b)

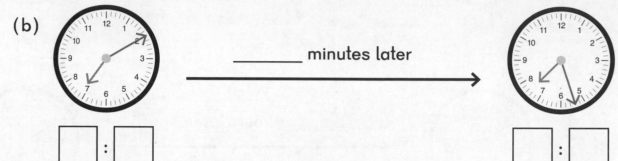

_____ minutes later

☐ : ☐   ☐ : ☐

(c)

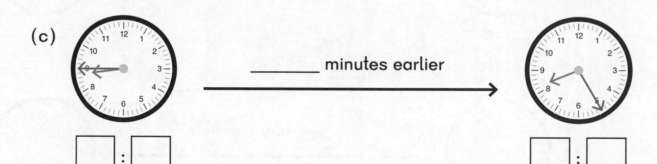

_____ minutes earlier

☐ : ☐   ☐ : ☐

(d)

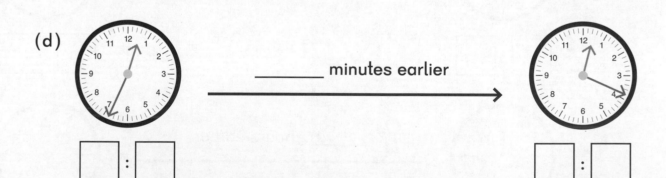

_____ minutes earlier

☐ : ☐   ☐ : ☐

3 Write the times and fill in the blanks.

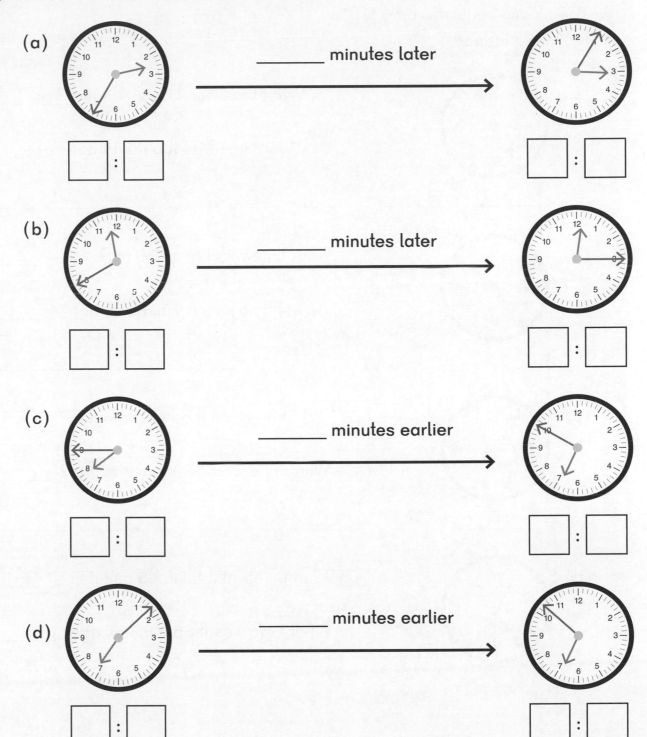

(a) _____ minutes later

□ : □   □ : □

(b) _____ minutes later

□ : □   □ : □

(c) _____ minutes earlier

□ : □   □ : □

(d) _____ minutes earlier

□ : □   □ : □

Practice

4 The clocks show the time.
Fill in the blanks.

(a)

What time will it be in 4 hours? _____

What time was it 5 hours ago? _____

(b)

What time will it be in 30 minutes? _____

What time was it 7 minutes ago? _____

(c)

What time will it be in 7 minutes? _____

What time was it 30 minutes ago? _____

(d)

What time will it be in 45 minutes? _____

What time was it 55 minutes ago? _____

5 Aliyah's piano lesson started at 3:30.
The lesson was 45 minutes long.
What time did it end?

6 A concert lasted 2 hours.
It ended at 11:40.
What time did it start?

7 A ferry goes from the mainland to an island.
The ferry arrived at the island at 6:10.
The crossing time was 42 minutes.
What time did the ferry leave the mainland?

Challenge

8 A ferry leaves at 10:55.

Manuel wants to be in line for the ferry an hour ahead of time.

It takes him 2 hours to drive from home to the ferry.

What time should he leave home?

9 Draw hands on the clock with the missing hands to complete the patterns.

(a)

(b)

Exercise 3

Basics

1 Fill in the blanks.

(a) 1 day = ☐ hours

(b) There are _____ hours between midnight and noon.

(c) There are _____ hours between noon and midnight.

2 Write a.m. or p.m. in the blanks.

(a)
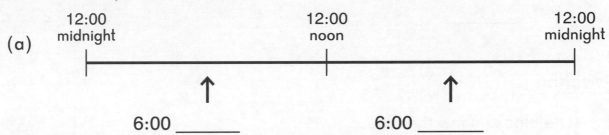

(b) 10:00 _____ is between noon and midnight.

(c) Jamal woke up after a night's sleep at 7:00 _____.

(d) Imani finished dinner at 6:15 _____.

(e) Evan ate lunch at 12:30 _____.

(f) A soccer game began at 11:30 _____ and ended at 12:45 _____.

(g) Kai wanted to finish all his gardening before noon, so he began at 9:00 _____.

(h) To see the night-time glowing algae in the water, Jo was on the beach at 10:00 _____.

3 Write the times using a.m. or p.m.

(a)

6 hours later ────────────────────────►

8:50 a.m.

_____

(b)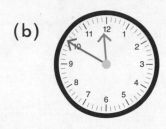

_____ minutes later ────────────────►

_____

12:25 a.m.

Practice

4 The clocks show the time.
Fill in the blanks, including a.m. or p.m.

(a)

What time will it be in 4 hours? _____

What time was it 40 minutes ago? _____

_____ a.m.

(b)

What time was it 4 hours ago? _____

What time will it be in 40 minutes? _____

_____ p.m.

5 Write the times, including a.m. or p.m.

(a) 5 minutes after 10 in the morning

☐ : ☐ _____

(b) Half past 1 in the afternoon

☐ : ☐ _____

(c) A quarter to 7 in the evening

☐ : ☐ _____

(d) 5 minutes to 10 in the morning

☐ : ☐ _____

(e) 23 minutes after midnight

☐ : ☐ _____

6 Raj's tennis lesson ended at 12:35 p.m.
The lesson was 50 minutes long.
What time did it start?

7 Cora went on a 3-hour hike.
The hike ended at 2:40 p.m.
What time did it start?

Challenge

8 There are 24 different time zones around the world.
The table below shows the times in some cities in the world when it is midnight in Honolulu.

Honolulu	12 midnight
Denver	4:00 a.m.
New York City	6:00 a.m.
London	11:00 a.m.
Moscow	1:00 p.m.
Singapore	6:00 p.m.
Auckland	10:00 p.m.

Auckland is 22 hours ahead of Honolulu, so when it is 12 midnight in Honolulu, it is 10 p.m. in Auckland.

(a) Singapore is _____ hours ahead of London.

(b) When it is 10:00 p.m. in London, it is _____ in Singapore.

(c) Honolulu is _____ hours behind New York City.

(d) When it is 10:00 p.m. in Honolulu, it is _____ in New York City.

(e) Moscow is _____ hours behind Auckland.

(f) When it is 4:10 a.m. in Moscow, it is _____ in Auckland.

(g) When it is 4:10 a.m. in Auckland, it is _____ in Moscow.

Exercise 4

Check

1 What time is it?

(a)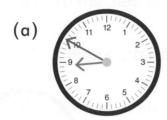

(b)

(c)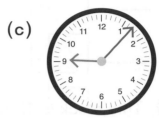

_____   _____   _____

2 Draw the minute hands to show the times.

(a)

(b)

(c)

   20 minutes to 4   a quarter past 1   6:18

3 (a) When the minute hand moves halfway around the clock,

_____ minutes have passed.

(b) When the hour hand moves halfway around the clock,

_____ hours have passed.

(c) When the minute hand moves a quarter of the way around the clock,

_____ minutes have passed.

(d) When the hour hand moves a quarter of the way around the clock,

_____ hours have passed.

Challenge

11 It is 12 hours later in Singapore than it is in New York City.

It is 6 hours earlier in Berlin than it is in Singapore.

If it is 10:00 a.m. in New York City, what time is it in Berlin?

12 It is 3 hours earlier in Honolulu than in Seattle.

An airplane flight from Seattle to Honolulu takes 6 hours.

(a) If it is 1:00 p.m. in Seattle, what time is it in Honolulu?

(b) A flight leaves Seattle at 1:00 p.m.

What time will it be in Honolulu when the flight lands there?

(c) Another flight from Seattle arrives in Honolulu at 1:00 p.m.

What time was it in Seattle when the flight took off from there?

13 Some countries do not use a.m. and p.m.

They use a 24-hour clock, instead of a 12-hour clock.

1:00 p.m. in 12-hour time is 13:00 in 24-hour time.

What time is it in 24-hour time at 6:00 p.m.?

Chapter 13 Capacity

Basics

1 The diagram shows how much water each container can hold.

Fill in the blanks.

(a) Container P can hold less water than Container _____
and more water than Container _____.

(b) Container _____ has the greatest capacity.

(c) Container _____ has the least capacity.

2 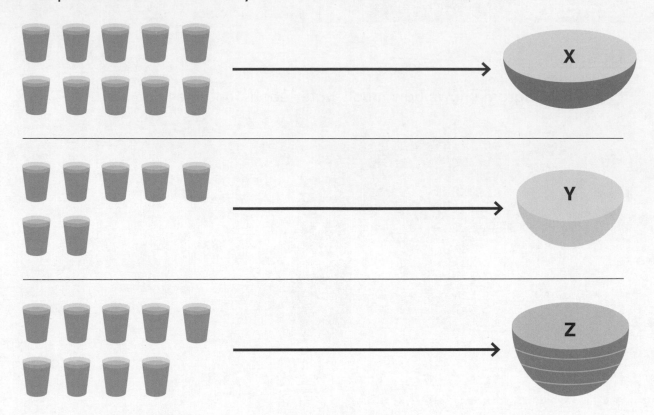 is 1 unit.

The picture shows how many units of water will fill the bowls.

(a) Bowl Y can hold _____ units of water.

(b) Bowl _____ can hold 10 units of water.

(c) Bowl _____ has the greatest capacity.

(d) Bowl _____ has the least capacity.

(e) Bowl X can hold _____ more units of water than Bowl Y.

(f) Bowl Y can hold 2 fewer units of water than Bowl _____.

(g) 16 units of water can completely fill bowls _____ and _____.

Practice

3 🥤 is 1 unit.

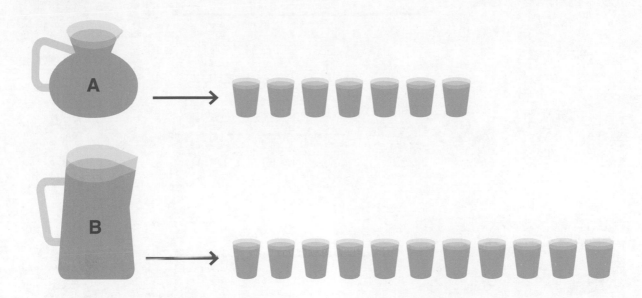

The total capacity of Containers A and B is _____ units.

4

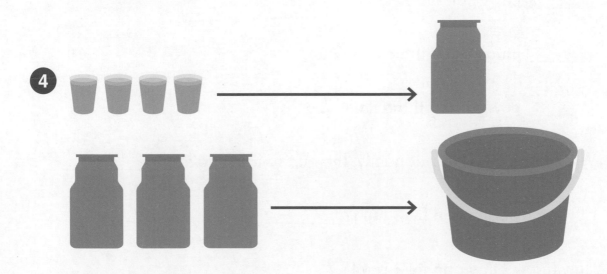

(a) The bucket can hold _____ bottles of water.

(b) The bucket can hold _____ cups of water.

(c) The bucket can hold _____ more cups of water than the bottle.

(d) 2 bottles and 1 bucket can hold _____ cups of water.

Challenge

5

(a) _____ cups can fill the jug.

(b) _____ mugs can fill the same jug.

(c) _____ bowls can fill the same jug.

(d) Which has a greater capacity, the cup or the mug? _____

(e) How many cups can fill 1 mug? _____

(f) How many cups can fill 1 bowl? _____

(g) 3 jugs can hold the same amount of water as _____ mugs.

(h) Can 1 jug hold enough water to fill 6 cups and 3 mugs? _____

(i) 3 jugs have the same capacity as 4 mugs and _____ bowls.

Exercise 2

Basics

1 Look for some containers you think hold about 1 liter.

Use a 1-liter container to measure their capacity.

Put a check (✓) in the correct box.

Container	Less than 1 L	1 L	More than 1 L

2 Look for some containers that hold more than 1 liter.

Estimate about how many liters of water they hold.

Then use the 1-liter container to find their capacity.

Container	Estimated	Measured
	About _____ L	Between _____ L and _____ L
	About _____ L	Between _____ L and _____ L
	About _____ L	Between _____ L and _____ L
	About _____ L	Between _____ L and _____ L

3

The _____ has the greatest capacity.

he bucket can hold _____ L.

e capacity of the pitcher is _____ L less than the bucket.

(d) The thermos can hold about _____ L.

(e) Altogether, the three containers can hold almost _____ L.

Practice

4 Circle the most reasonable capacity for each of the following.

(a) A fish tank in the home.

1 L   40 L   800 L

(b) A bottle of juice.

1 L   10 L   50 L

(c) A bath tub.

20 L   200 L   900 L

5 A water tank can hold 400 L of water.
It has 132 L of water in it now.
How much more water is needed to fill the tank?

6 A pot has 36 L of broth.
Ms. Ivanov ladled 4 L of broth into each jar.
How many jars did she use?

Challenge

7

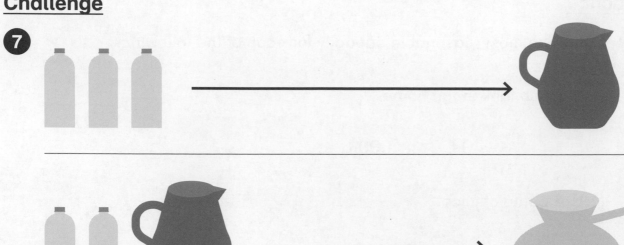

Each water bottle can hold 1 L of water.
How many liters can the bucket hold?

8 Using only these three containers, how can you divide the water in the 12-L container evenly between all 3 containers?

12 L   9 L   4 L

Check

1

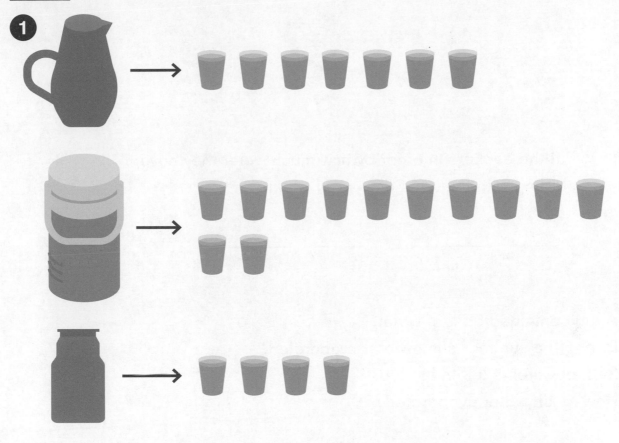

(a) The _____ has the least capacity.

(b) The pitcher has a greater capacity than the _____.

(c) The pitcher can hold _____ cups more than the bottle.

(d) The total capacity of the pitcher, thermos, and bottle is _____ cups.

(e) The thermos could be filled up using the bottle _____ times.

(f) 4 pitchers can fill _____ cups.

(g) 19 cups can completely fill the _____ and
the _____.

2 (a) Which beaker contains 1 L of water? _____

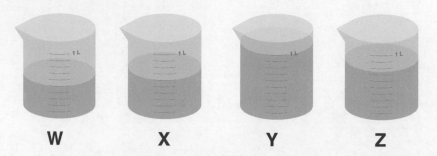

W   X   Y   Z

(b) Put the beakers in order by how much water they have, starting with the least amount of water.

_____   _____   _____   _____

3 A tub contained 55 L of water.
During the summer some water evaporated.
21 L of water is left in the tub.
How much water evaporated?

4 A fish tank can hold 45 L of water.
A bucket can hold 5 L of water.
How many full buckets of water are needed to fill the tank?

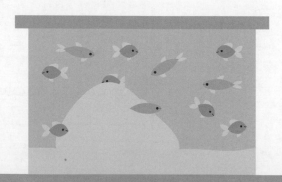

5 Ximena drank about 3 L of water each day.
About how many liters of water did she drink in a week?

6 A small hot water tank can hold 225 L of water.
A large hot water tank can hold 320 L of water.
How much more water can the large tank hold than the small tank?

7 A quart is a little less than 1 liter.
Which is more, 10 liters or 10 quarts?

8 4 quarts of water is the same amount of water as 1 gallon of water.

(a) How many quarts of water are needed to fill a 5-gallon tank?

(b) How many 1-gallon tanks can be filled with 40 quarts of water?

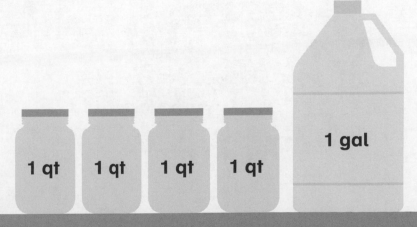

1 qt   1 qt   1 qt   1 qt   1 gal

Challenge

9 Each can holds 5 L of paint.
Santino mixed 6 cans of white paint and 3 cans
of red paint to get pink paint.
How much pink paint does he have?

10 A bucket can hold 12 L of water.
A pail can hold 3 L of water.
How many more liters of water can 2 buckets hold than 5 pails?

11 How can you use these containers to put 11 L of water from the
tub into the 12-liter jug using the fewest number of steps?

13-3 Practice

Chapter 14 Graphs

Basics

1 Members of a club voted to choose the color for their new club t-shirt. This picture graph shows the results.

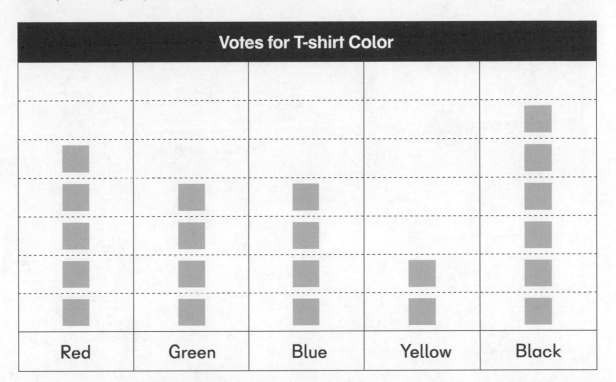

Each ▨ stands for 5 votes.

(a) _____ got the most votes.

(b) _____ club members voted for yellow.

(c) _____ more club members voted for red than for yellow.

(d) _____ fewer club members voted for blue than for black.

(e) The colors _____ and _____ received the same number of votes.

Practice

2 Count the number of each kind of bug.
Then color the circles in to complete the graph on the next page.

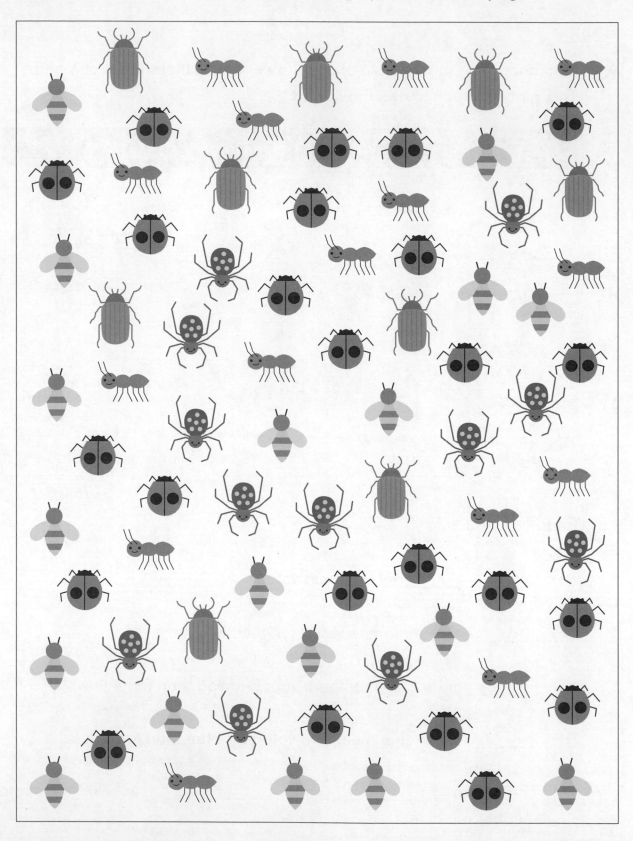

14-1 Picture Graphs

Bugs

Beetle	Ant	Ladybug	Bee	Spider
○	○	○	○	○
○	○	○	○	○
○	○	○	○	○
○	○	○	○	○
○	○	○	○	○
○	○	○	○	○
○	○	○	○	○
○	○	○	○	○

Each ● stands for 3 bugs.

(a) There are _____ ladybugs.

(b) There are _____ more spiders than beetles.

(c) There are _____ fewer ants than ladybugs.

(d) There are _____ more ladybugs and bees than beetles and ants.

(e) Put the bugs in order of least to greatest number counted.

_____, _____, _____, _____, _____

3 This picture graph shows which fruit some children chose for snack.

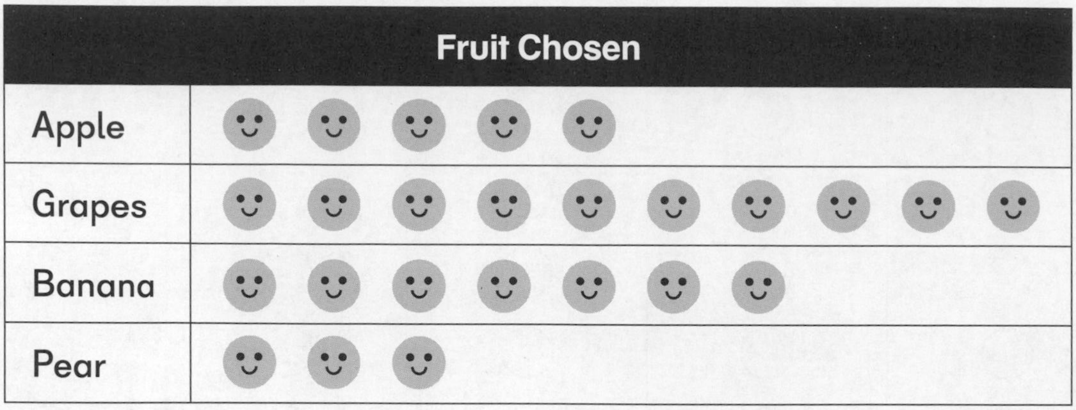

Fruit Chosen										
Apple	☺	☺	☺	☺	☺					
Grapes	☺	☺	☺	☺	☺	☺	☺	☺	☺	☺
Banana	☺	☺	☺	☺	☺	☺	☺			
Pear	☺	☺	☺							

Each ☺ stands for 2 children.

(a) The children chose _____ the most.

(b) _____ children chose bananas.

(c) _____ fewer children chose pears than apples.

(d) The graph shows the choices of _____ children.

4 Each ◆ stands for 4 cars.

(a) ◆◆◆◆ stands for _____ cars.

(b) Color the correct number of shapes to show 24 cars.

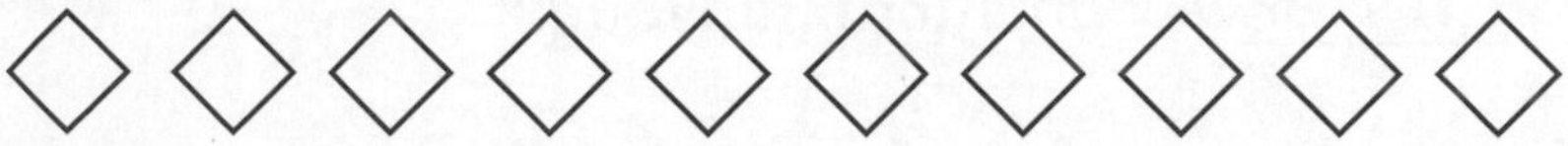

◇ ◇ ◇ ◇ ◇ ◇ ◇ ◇ ◇ ◇

5 ▲▲▲▲▲ stands for 50 plants.

▲▲ stands for _____ plants.

Basics

1 This bar graph shows the number of times certain pizza toppings were ordered one day at a pizzeria.

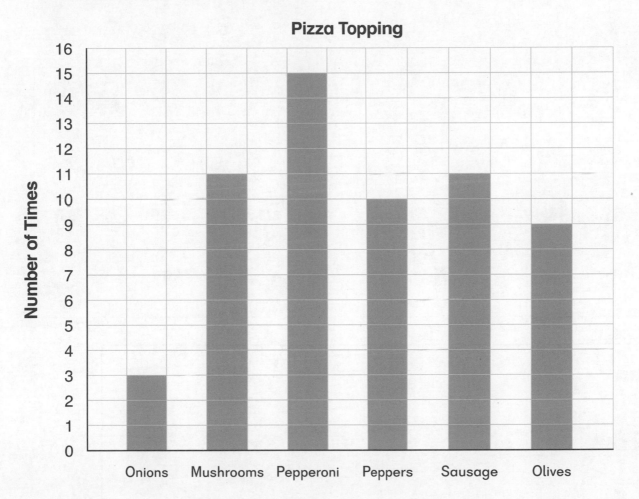

(a) _____ was ordered the most.

(b) _____ were ordered the least.

(c) Olives were ordered _____ times.

(d) Peppers were ordered _____ fewer times than pepperoni.

(e) _____ and _____ were ordered the same number of times.

Practice

2 Count the number of each kind of sticker (circle, moon, star, square, and triangle).

Complete the bar graph.

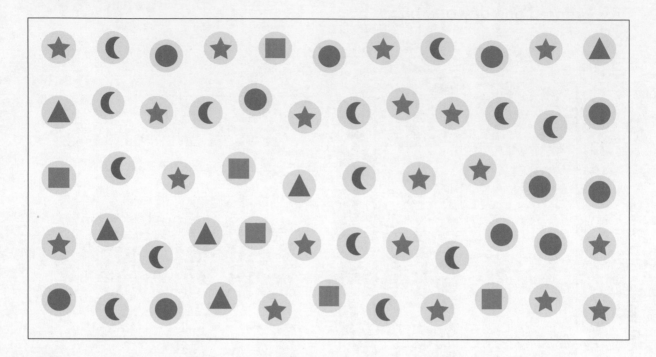

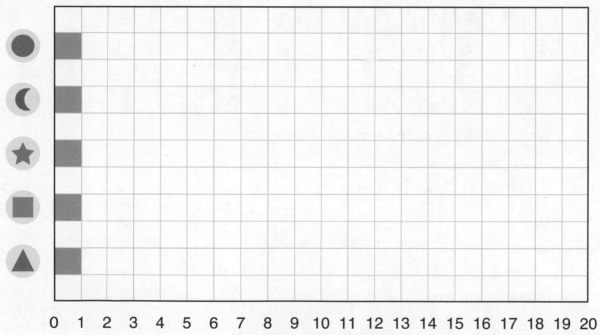

Number of Stickers

14-2 Bar Graphs

3 Complete the bar graph below using the information from this picture graph.

Fruit Chosen	
Apple	☺ ☺ ☺ ☺ ☺
Grapes	☺ ☺ ☺ ☺ ☺ ☺ ☺ ☺ ☺ ☺
Banana	☺ ☺ ☺ ☺ ☺ ☺ ☺
Pear	☺ ☺ ☺

Each ☺ stands for 2 children.

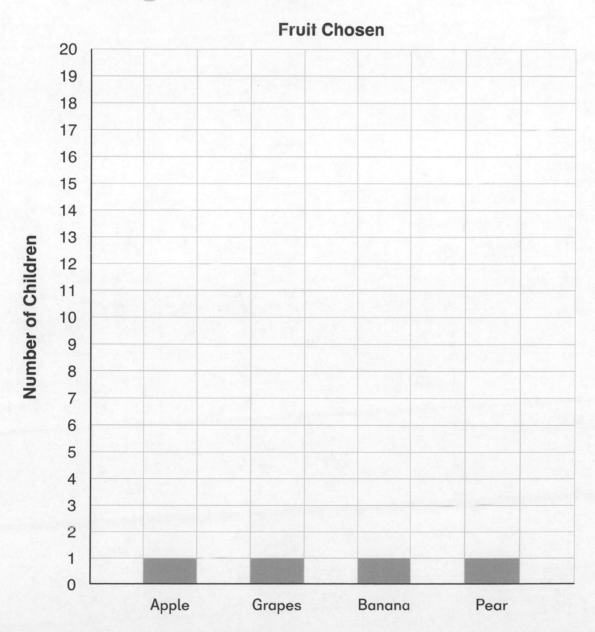

Fruit Chosen

Challenge

4 Use the information below to complete the bar graph.

Riya has 13 stamps.

Taylor has 9 more stamps than Riya.

Pedro has 6 fewer stamps than Taylor.

Liam has the same number of stamps as Pedro.

Altogether, the five children have 80 stamps.

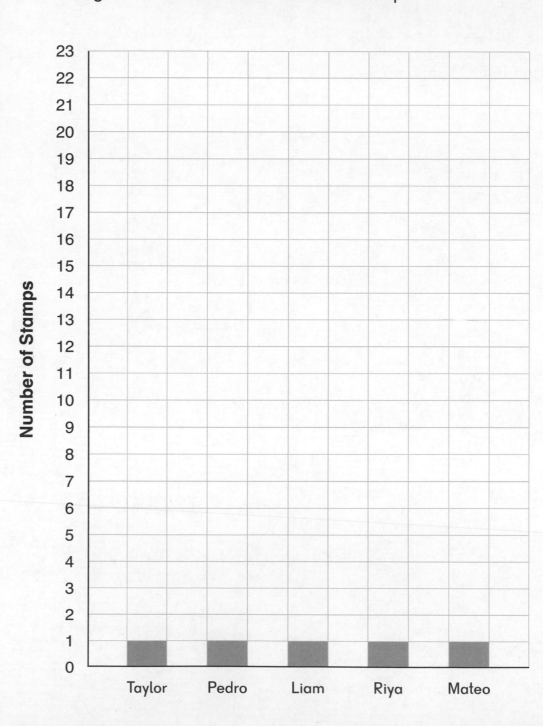

Check

1 This table shows the number of bottles collected by some children for a recycling project.

Bottles Collected	
Diego	🍼🍼🍼🍼🍼🍼🍼
Holly	🍼🍼🍼🍼🍼🍼🍼🍼🍼
Susma	🍼🍼🍼🍼🍼🍼
Micah	🍼🍼🍼🍼
Noah	🍼🍼🍼🍼🍼🍼🍼

Each 🍼 stands for 3 bottles.

(a) _____ collected the most bottles.

(b) Diego collected _____ bottles.

(c) Holly collected _____ more bottles than Noah.

(d) Susma collected 6 fewer bottles than _____.

(e) Micah collected more bottles than is recorded on the graph.
He collected a total of 24 bottles.
How many more 🍼 should be drawn on the graph for Micah's bottles?

2 This table shows the number of four types of fish sold at a fish store.

Fish	Goldfish	Swordtail	Guppy	Angelfish
Number	40	35	25	15

(a) Use this information to complete the picture graph below.

Fish Sold				
Goldfish	Swordtail	Guppy	Angelfish	Clownfish

Each ⊂×| stands for 5 fish.

(b) 15 fewer clownfish were sold then swordtail.
Draw ⊂×| for the number of clownfish sold.

(c) How many of the 5 types of fish were sold in all?

(d) If the angelfish and clownfish cost $3 each, how much more money was collected from selling clownfish than angelfish?

3 This table is a tally for the colors of some cars in a parking lot.

Red	卌 ////
Gray	卌 卌 ////
Black	卌 卌 //
White	卌 卌

(a) Use the information to complete the bar graph.

Car Color

Number of Cars

Red   Gray   Black   White

(b) There were 4 more blue cars than red cars and 8 fewer green cars than blue cars.
Add this information to the graph.

(c) _____ cars were counted altogether.

(d) Put the car colors in order from greatest to least number counted.

_____, _____, _____, _____, _____, _____

Challenge

 4 This picture graph shows the number of burritos sold by some children at a fund raiser last week.

Use the information below to write the correct name in each column.

Matias sold 32 burritos.

Paula sold 20 burritos fewer than Matias.

Misha sold 4 more burritos than Paula.

Nolan sold twice as many burritos as Paula.

Each burrito cost $3.

Sasha raised $24 selling burritos.

Complete the graph for the number of burritos Sasha sold.

Burritos Sold				
	🌯			
	🌯			
🌯	🌯			
🌯	🌯			
🌯	🌯	🌯		
🌯	🌯	🌯	🌯	
🌯	🌯	🌯	🌯	
🌯	🌯	🌯	🌯	
				Sasha

Each 🌯 stands for 4 burritos.

Chapter 15 Shapes

Basics

1 Check the figures that are made from only curved lines.

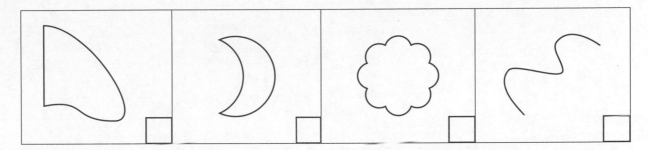

2 Check the figures that are made from only straight lines.

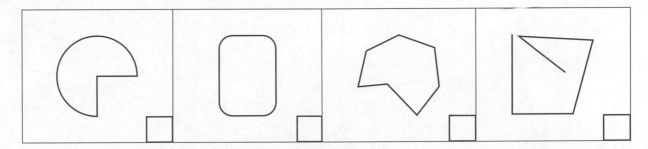

3 Check the figures that are closed shapes.

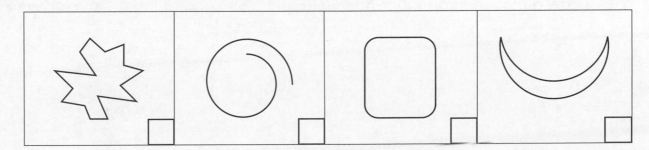

Practice

4 Use a ruler to draw only straight lines to make each of these figures closed shapes.

(a)

(b)

(c)

5 Draw only curved lines to make each of these figures closed shapes.

(a)

(b)

(c)

6 Draw a closed shape that has 4 straight sides and at least 1 curved side.

Exercise 2

Basics

1 Check the figures that are polygons.

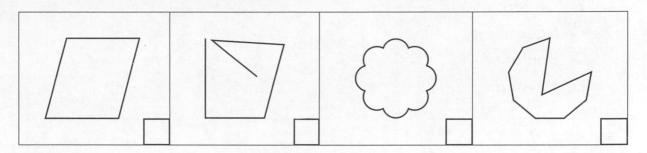

2 Check the figures that are quadrilaterals.

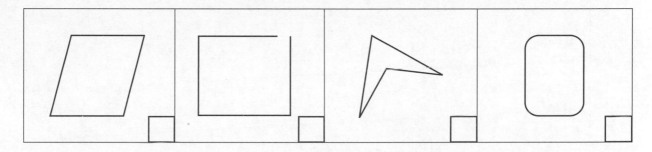

3 (a) A hexagon has _____ sides and _____ corners.

(b) A pentagon has _____ sides and _____ corners.

(c) A polygon with 20 sides has _____ corners.

(d) A polygon has _____ curved sides.

(e) The fewest number of sides a polygon can have is _____.

Practice

 Use the dot paper and a ruler to draw two of each of the following shapes.

A Triangle

B Quadrilateral

C Square

D Pentagon

E Hexagon

5 Draw straight lines on each figure to divide it into the given shapes.

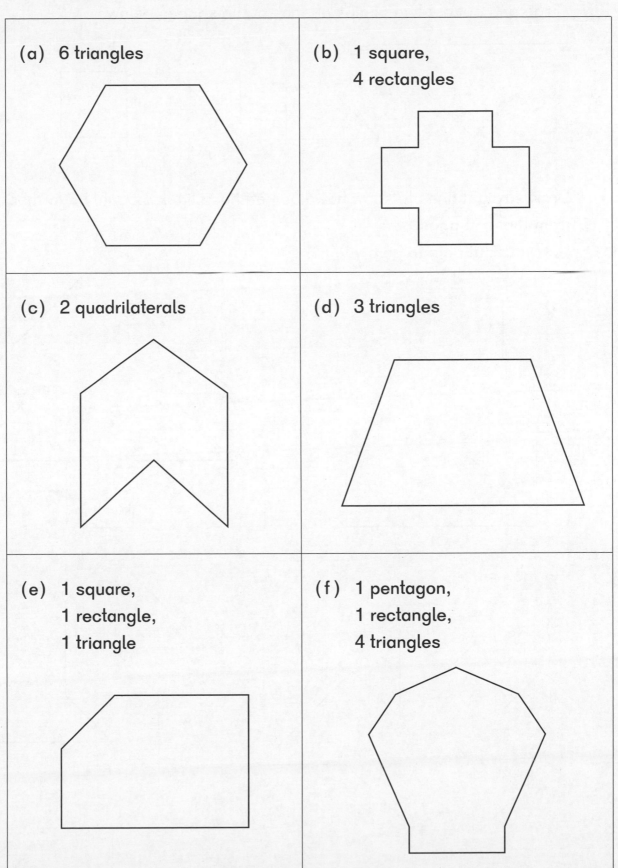

(a) 6 triangles

(b) 1 square,
4 rectangles

(c) 2 quadrilaterals

(d) 3 triangles

(e) 1 square,
1 rectangle,
1 triangle

(f) 1 pentagon,
1 rectangle,
4 triangles

Challenge

6 Trace and cut out two copies of each of the shapes below.

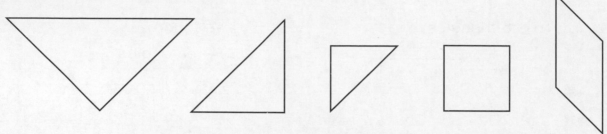

Draw straight lines to show how each of these figures could be formed from these shapes.
Use your cut outs to help you.

(a)

(b)

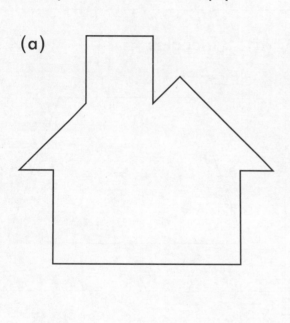

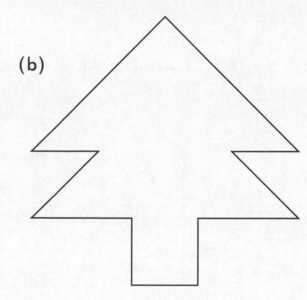

(c)

(d)

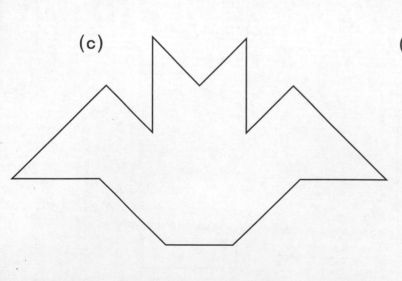

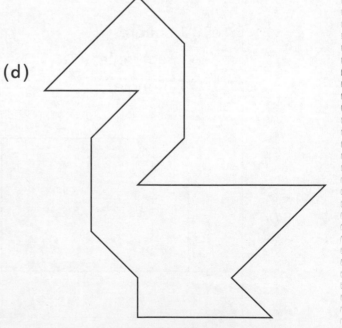

15-2 Polygons

Basics

1 Use a ruler to draw a straight line to divide each circle into semicircles.

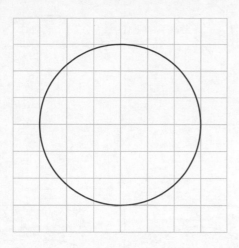 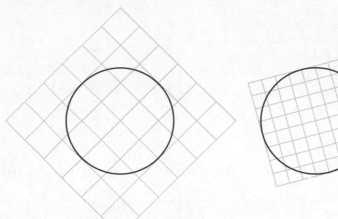

2 Use a ruler to draw straight lines to divide each circle into quarter-circles.

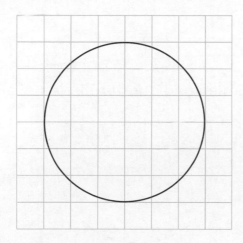 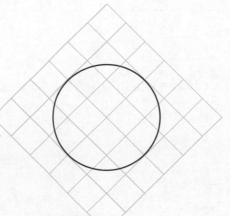 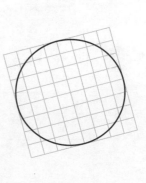

3 (a) A semicircle has _____ curved side and _____ straight side.

(b) There are _____ semicircles in a circle.

(c) A quarter-circle has _____ curved side and _____ straight sides.

(d) There are _____ quarter-circles in a circle.

Practice

4 Check the figures that are semicircles.

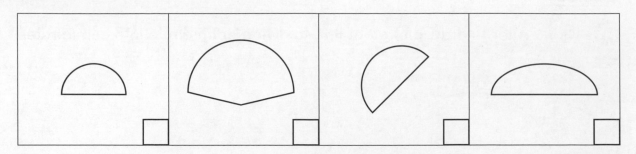

5 Check the figures that are quarter-circles.

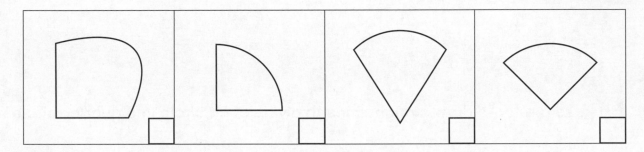

6 Draw lines to divide each figure into semicircles and quarter-circles.

(a)

(b)

(c)

(d)

7 Draw straight lines on each figure to divide it into the given shapes.

(a) 1 semicircle,
1 rectangle,
1 triangle

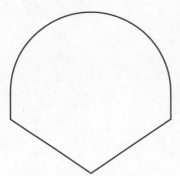

(b) 1 square,
1 triangle,
3 quarter-circles

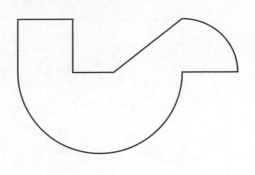

8 Fill in the blanks with the names of the shapes:
quarter-circle, semicircle, triangle, or rectangle.

(a) This figure is made from

2 _____,

a _____,

and a _____.

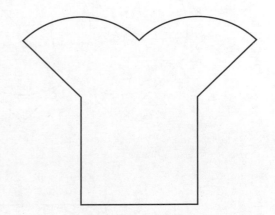

(b) This figure is made from

a _____,

a _____,

and a _____.

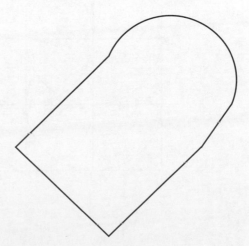

Challenge

9 How many quarter-circles are shown in each drawing?

(a)

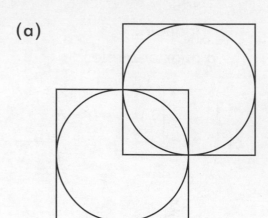

(b)

(c)

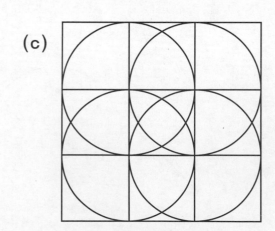

Basics

1 Study each pattern.

Underline the shapes that are repeated (the first one is done for you.)

Circle the shape that comes next.

(a)

Circle what changes.

shape   **size**   **shade**

(b)

Circle what changes.

shape   **orientation**   **shade**

(c)

Circle what changes.

shape   **orientation**   **shade**

Practice

2 Circle the figure that is missing in each pattern.

(a)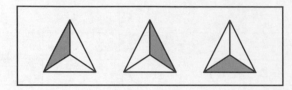

(b)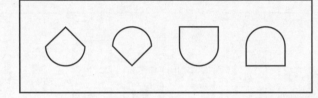

3 Draw the missing figure in each pattern.

(a)

(b)

(c)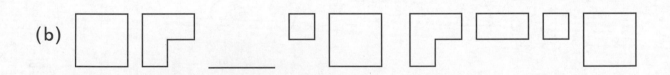

4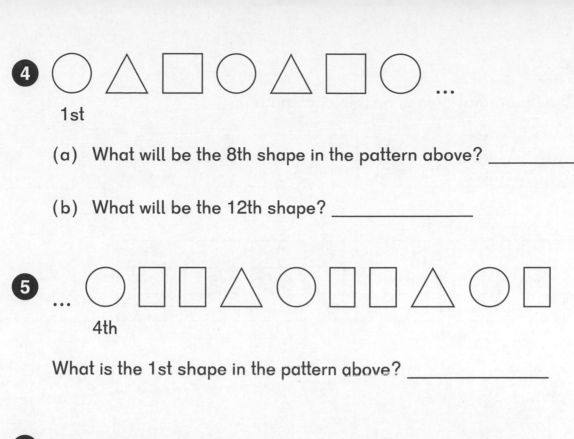

1st

(a) What will be the 8th shape in the pattern above? _____

(b) What will be the 12th shape? _____

5 ...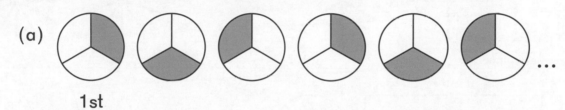

4th

What is the 1st shape in the pattern above? _____

6 Shade the shapes to match the shape in the given position for each pattern.

(a)

1st

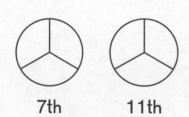

7th   11th

(b)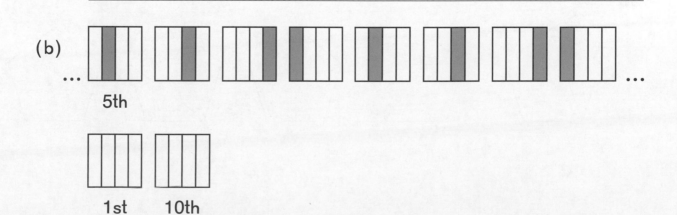

5th

1st   10th

Challenge

 7 Draw the figure that comes next in each pattern.

(a)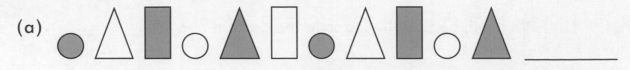

(b)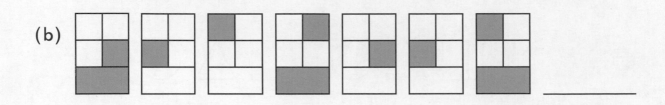

Draw the missing figures.

(a)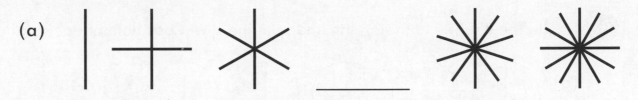

(b)

9 Draw the arrow in the pentagon to show what comes next.

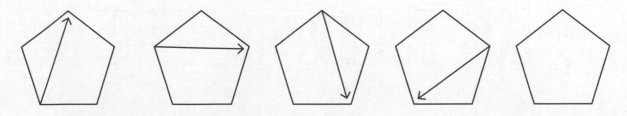

Basics

1 Match similar solids.

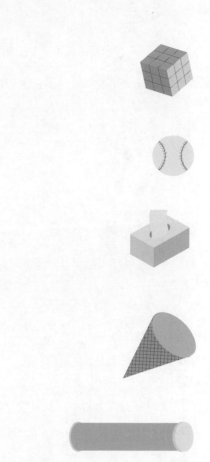

2 Cross out the solid that does not belong.

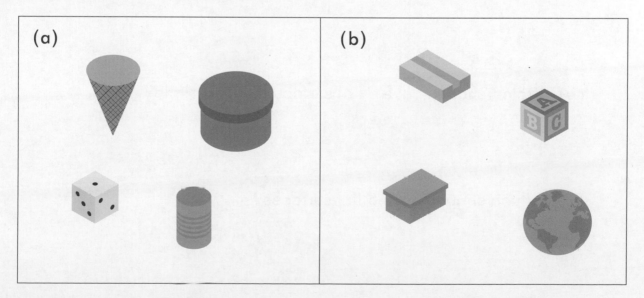

(a)

(b)

3 Match.

sphere

cylinder

cuboid

cone

cube

Practice

4 Answer with cuboid, cube, cylinder, sphere, or cone.

(a) Which shapes have 1 curved surface?

(b) Which shapes have a flat surface that is a circle?

(c) Which shapes have 6 flat surfaces?

5 Name the shape of the shaded faces.

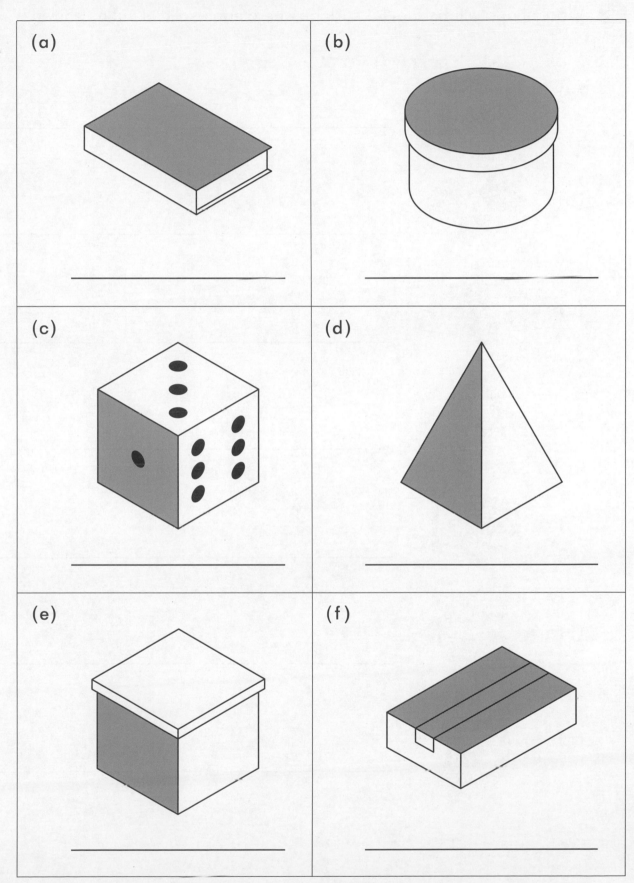

(a)

_____

(b)

_____

(c)

_____

(d)

_____

(e)

_____

(f)

_____

Challenge

6 What shape will be the flat surface where each solid is cut as shown?

(a)  _____

(b)  _____

(c)  _____

(d)  _____

(e)  _____

(f)  _____

(g) 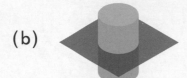 _____

Exercise 6

Check

1 Write the number of straight lines in each of the following figures.
Circle the figures that are closed.
Cross out the polygon.

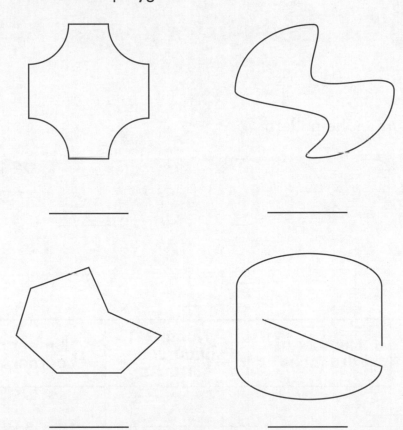

_____

_____

_____

_____

2 Draw straight lines to divide each figure into the fewest possible
quarter-circles, half-circles, quadrilaterals, and triangles.

(a) 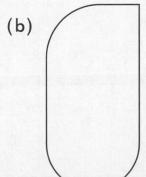   (b)   (c)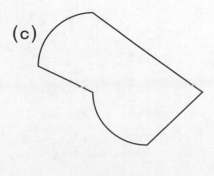

3 Write the name of the shape that comes next in each pattern.

(a) 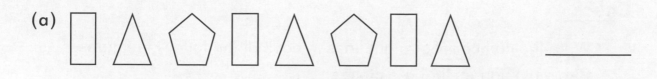 _____

_____

(b)  _____

4 Draw the missing figure in the pattern.

_____

5 Complete the table.

Solid	Number of flat faces	Number of curved surfaces	Number of corners
Cuboid			
Cone			
Cylinder			
Sphere			

6 Use a ruler to copy the figures onto the dot grid.
Then draw straight lines to divide each of your copied figures into triangles.

(a)

(b)

(c)

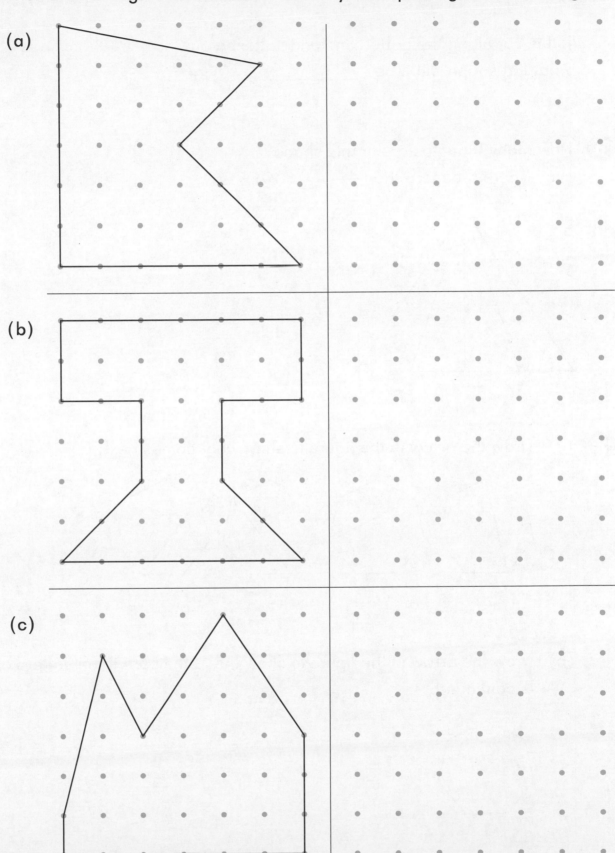

Challenge

7 A paper roll cardboard is shaped like a _____.

If it is cut along the dotted line and flattened out,
what flat shape will it be? _____

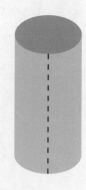

8 How many triangles are in this shape?
Hint: There are more than 4.

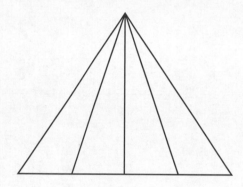

9 (a) Draw the arrow in the figure to show what comes next.

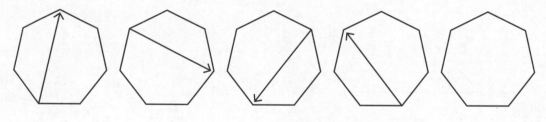

1st

(b) Draw the arrow in the figure to show the 10th figure if the pattern above
is continued.

Check

| 680 | 829 | 983 | 806 | 68 |

Use the numbers above for each of the following problems.

(a) The numbers with 8 in the tens place are _____.

(b) The numbers with 8 in the hundreds place are _____.

(c) The digit in the hundreds place in 893 is _____ more than the digit in the ones place.

(d) The digit in the ones place in _____ is 2 more than the digit in the ones place in _____.

(e) Write the numbers in order from least to greatest.

(f) 30 more than _____ is 859.

(g) 200 less than _____ is 782.

(h) 70 more than _____ is 750.

(i) 6 less than _____ is 977.

(j) The digits in _____ add to 20.

2 Write the missing digits.

(a)
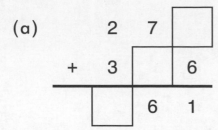

```
      2  7  ☐
  +   3  ☐  6
   ☐  6  1
```

(b)
```
   ☐  3  0
 -    2  7  ☐
      6  ☐  2
```

3 What fraction of each shape is shaded and what is the name of the shape?

(a)

_____ of the _____
is shaded.

(b)

_____ of the _____
is shaded.

4 (a)

The time is _____ minutes
past _____.

(b)

The time is _____ minutes
to _____.

5 How much weight needs to be added to balance the scale?

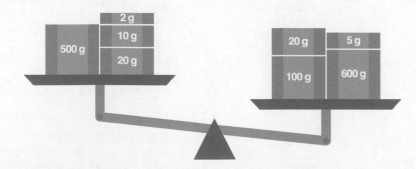

_____ g needs to be added to the left side.

6 Use a centimeter ruler to measure all three sides of this triangle.
The total length of all three sides is _____ cm.

7 What is the total amount of money?

$_____._____

8 5 equal size buckets can hold 40 L of water.

(a) How much water can 1 of these buckets hold?

(b) 3 buckets are needed to fill a barrel.
What is the capacity of the barrel?

9 There were 389 adults and 562 children at an amusement park.
How many fewer adults were there than children?

10 Chapa saved $3.85.
Isabella saved $5.35.

(a) How much more did Isabella save than Chapa?

(b) How much did they save altogether?

11 A notebook cost $2.45.
A binder costs 5 quarters and 1 dime more than the notebook.
How much does the binder cost?

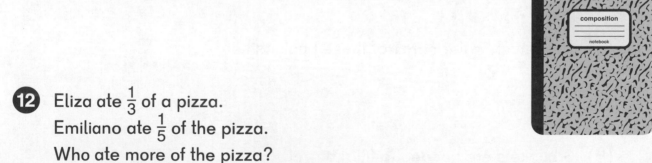

12 Eliza ate $\frac{1}{3}$ of a pizza.
Emiliano ate $\frac{1}{5}$ of the pizza.
Who ate more of the pizza?

13 This picture graph shows the number of some animals in a pet store.

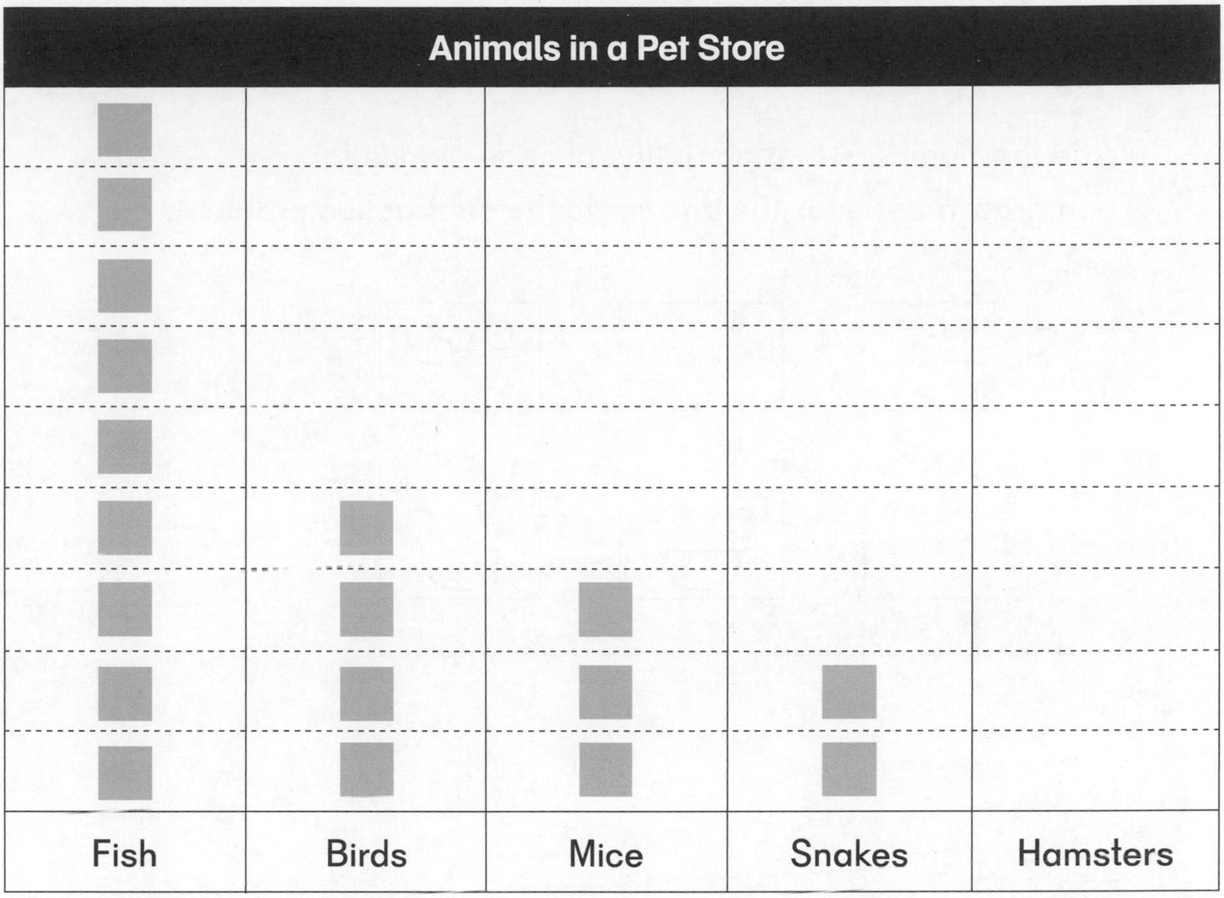

Animals in a Pet Store				
Fish	Birds	Mice	Snakes	Hamsters

There are 15 mice and snakes altogether.

(a) Each ▦ stands for _____ animals.

(b) How many of each animal are there?

 Fish _____   Birds _____   Mice _____   Snakes _____

(c) There are 18 hamsters. Draw the correct number of ▦ for the hamsters.

(d) The pet store got more fish and now has twice as many fish. How many fish are there now? _____

(e) How many of all 5 kinds of animals are there altogether now? _____

Challenge

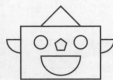   is a code for 472.

Write the numbers that each line of code stands for,
Then draw a code for the answer to the subtraction problem.

⑮ Circle the missing piece.

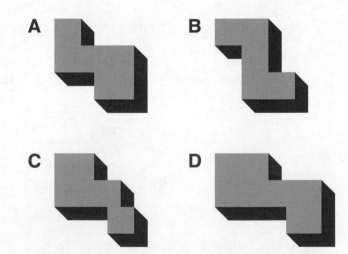

Exercise 8

Check

1 Write >, <, or = in the ◯.

(a) $500 + 40 + 3$ ◯ $30 + 400 + 5$

(b) $14 + 36 + 25$ ◯ $27 + 29 + 22$

(c) $580 + 70$ ◯ $710 - 60$

(d) $\$8.25 - \$1.80$ ◯ $\$6.25 + \$1.80$

(e) $\$4.30$ ◯ $403¢$

(f) $\frac{1}{5}$ ◯ $\frac{1}{8}$

(g) 7×4 ◯ 3×9

(h) $35 \div 5$ ◯ $32 \div 4$

(i) 1 m ◯ 100 cm

(j) $510\text{ g} + 480\text{ g}$ ◯ 1 kg

2 (a) $66 + \boxed{} = 100$

(b) $\boxed{} - 99 = 484$

(c) $\boxed{} + 8 = 322$

(d) $300 - 64 = \boxed{}$

(e) $45 \div \boxed{} = 9$

(f) $4 \times \boxed{} = 28$

(g) $62 + 24 + 30 + 15 = \boxed{} - 97$

3 (a) $\frac{3}{8}$ and _____ make 1.

(b) _____ and $\frac{7}{12}$ make 1.

4 How many rectangular faces are on each of these solids?

(a)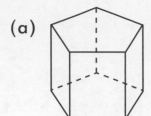

(b)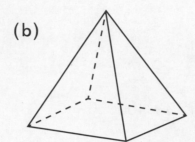

(c)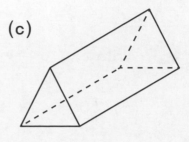

5 This clock shows the time Alex began doing his chores on Saturday.
Write the times using a.m. or p.m.

(a) What time did he begin his chores?

(b) He finished his chores 35 minutes later.
What time did he finish his chores?

(c) Then, he went outside to play for 4 hours.
What time did he finish playing?

6

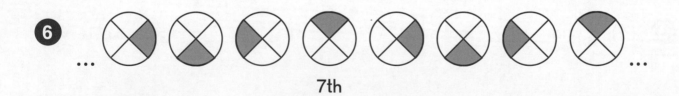

7th

Color the shape to show the shape in the given position in the pattern.

1st   3rd   12th   18th

7 A tank contains 132 L of water.
389 L are needed to fill it.
What is the capacity of the tank?

8 A bag of rice weighs 4 kg.
The price for 1 bag is $5.
Mr. Baker bought some rice for $25.

(a) How many bags of rice did he buy?

(b) How many kilograms of rice does he have?

9 Jasmine cut a cake into 10 equal pieces.
She and her 2 friends each ate a piece.

(a) What fraction of the cake did they eat?

(b) What fraction of the cake is left?

10 Circle the figure that completes the design.

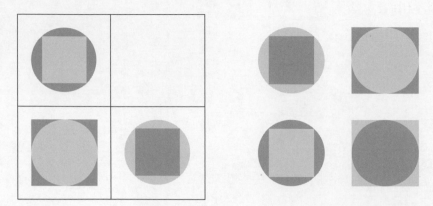

11 Use the information below to find the missing lengths.

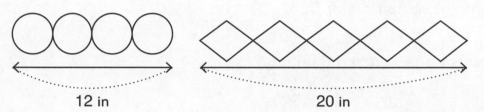

12 in   20 in

(a)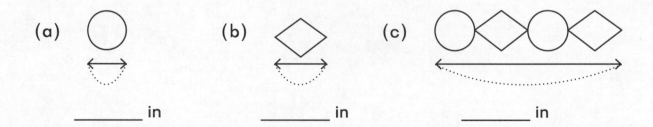

_____ in   _____ in   _____ in

(b)

(c)

(d)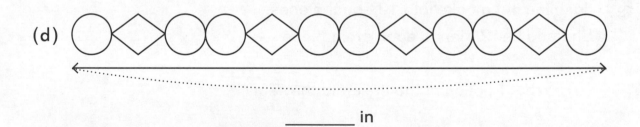

_____ in

(e) If 2 more shapes are added to continue the pattern in (d) above, what would be the new length?

12 (a) Use the information below to complete the bar graph.

$2.25 in quarters.

$1.70 in dimes.

The same number of nickels as dimes.

12 pennies.

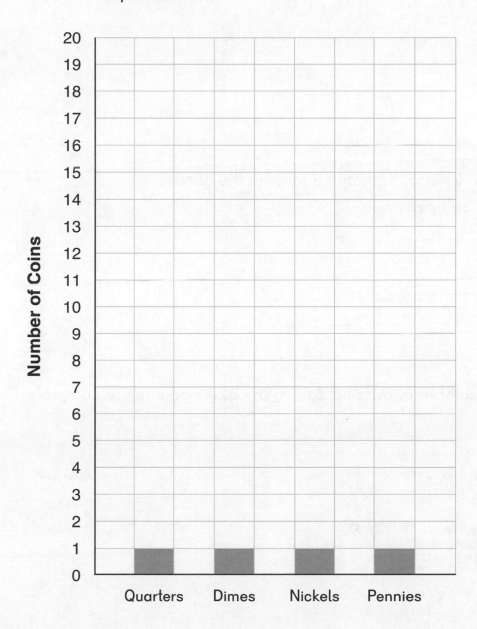

(b) The total number of coins is _____.

(c) The nickels make $_____

Challenge

13 Pablo cut a pie into fourths and put one piece on a plate.

He then cut that piece into thirds and ate one of those pieces.

What fraction of the pie is left?

Hint: Draw a picture.

14 Use the digits 0, 1, 2, 4, 5, 7, 8, and 9 to fill in the boxes.

Each number can be used once.

$$\square \times \square = \square\,\square$$

$$\square + \square = \square\,\square$$

15 Draw the correct number of missing dots in the blank squares to complete the puzzle.